Two Manuals Combined into One Book

Marine Scouting and Patrolling

plus

Soldier's Handbook for Individual Operations Survival in Cold Weather Areas

Department of Defense

Pentagon
Publishing

TABLE OF CONTENTS

PART I. SCOUTING

Chapter 7. Scouting Fire Team

PART 2. INFANTRY PATROLLING

Chapter 8. Fundamentals of Infantry Patrolling

Chapter 9. Patrol Organization

Chapter 10. Patrol Preparation

Chapter 11. Movement to and Return from the Objective Area

Chapter 12. Reconnaissance Patrols

Part I. Scouting

Chapter 1. Fundamentals of Scouting

Scouting involves observing terrain and/or the enemy, and accurately reporting those observations. Scouting requires proficiency in the use of weapons, cover and concealment, route selection, and skill in unobserved day or night movement.

1001. PURPOSE

When an infantry unit is not actively fighting the enemy, it should be actively searching for the enemy, which is the primary purpose of scouting. The unit attempts to keep the enemy off balance while making preparations for further attacks. Physically locating and keeping the enemy off balance are normally accomplished by small units ranging from a two-person scouting party to a squad-size patrol.

Infantrymen are sent out as scouts or as members of a patrol because the commander needs information about the enemy, terrain, and the location of friendly troops. The lives of the entire unit may depend upon the success or failure of a scout or patrol and the accuracy and timeliness of the report. The success of the scout or patrol will depend upon their training, preparation by the commander, and understanding their mission and the commander's requirements.

To wage combat successfully, a commander must have accurate, detailed, and timely information about the enemy, the terrain, and adjacent friendly units. Well-trained scouts and capably led patrols are among the most effective means the commander has for acquiring the information necessary to plan tactical actions and make decisions in execution.

1002. REQUIRED SCOUTING SKILLS

To be effective, a scout must be able to—

- Recognize terrain features.
- Read a map and determine direction.
- Practice and implement the principles of cover and concealment.
- Fully utilize movement and route selection.
- Know the enemy (estimate enemy unit composition and strength).
- Observe and report information accurately.
- Select routes and move through numerous types of terrain.

CHAPTER 2. TERRAIN, MAPS, AND DIRECTION

A scout must understand map symbols, identify elevations from contour interval lines, scale distance on a map, relate natural and man-made features shown on the map to the actual features on the ground, plot a course from one point to another, and locate his current position. To relate a map to the actual terrain and its features, a scout must be able to orient it to the ground using a compass, two points, a watch, and the sun or the stars.

2001. TERRAIN FEATURES

Since the infantry works and fights on the ground, terrain ashore information that scouts gather and report on is of great importance. Hills, valleys, woods, and streams are the forms and growths commonly referred to as natural land features. Artificial or man-made features include houses, bridges, and railroads. Figure 2-1 shows some important terrain features.

2002. THE LENSATIC COMPASS

The best method of finding direction, during both day and night, is with a compass. The lensatic and M2 are the two types of compasses issued to infantry units. Both work on the same principle.

The standard compass for general use in the Marine Corps is the pivot-mounted lensatic compass, so called because azimuths are read through a magnifying lens in the eyepiece. Figure 2-2, on page 2-2, shows the lensatic compass and its nomenclature. The meter graphic scale on the side of the compass is 1:50,000, which is the most used scale in military mapping. The graphic scale is useful in the field as a straightedge, an aid in orienting the map, and a means of reading map grid coordinates. The plastic dial is graduated in both degrees and mils. Numbers on the dial are printed in black. There are luminous markings on the bezel, floating dial, and on both ends of the sighting wire, plus a 3-degree bezel serration and clicking device that permit reading azimuths at night. The compass is carried in a nylon case that may be attached to the cartridge belt.

Magnetic compasses are affected by the presence of iron or magnetic fields. Consequently, the scout should not be within the influence of local magnetic

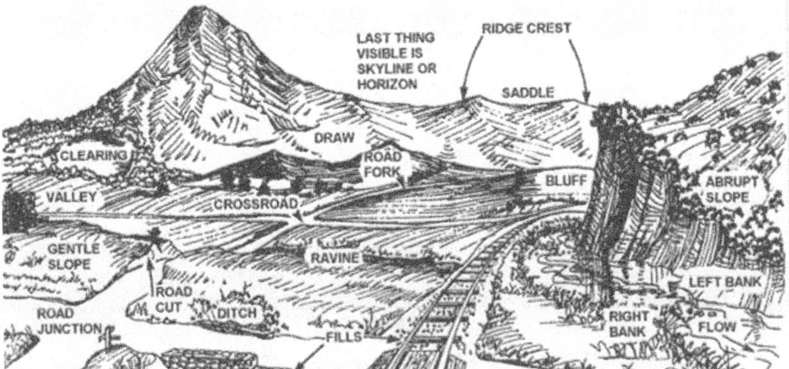

Figure 2-1. Natural and Artificial Terrain Features.

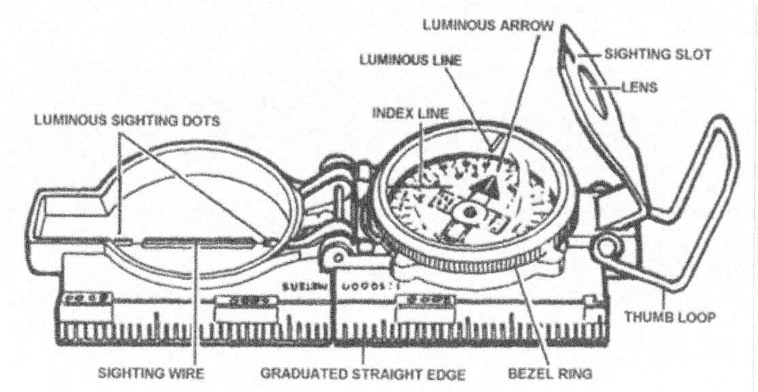

Figure 2-2. Lensatic Compass.

attraction while using a compass to determine direction. The rifle, pistol, and other metal objects must be laid aside when reading the compass. The minimum distances the scout should be from visible masses of iron and electrical fields of magnetism for the compass to provide accurate readings are provided below.

Magnetic Fields	Minimum Distances (meters)
High tension powerlines	60
Field guns	20
Vehicles (wheeled or tracked)	20
Telephone and telegraph wires	10
Barbed wire	10
Machine gun	3
Rifle, pistol	1

Reading an Azimuth

To read an azimuth to any point, the cover of the compass is raised to an angle of 90 degrees in relation to the index face, and the eyepiece is lifted to a 45-degree angle in relation to the bezel, or so the numbers on the dial can be seen. The thumb of either hand is placed in the thumb loop, the index finger extended along the side of the compass case, and the remainder of the hand closed. The closed hand and wrist are grasped with the other hand. The elbows are drawn in close to the body, forming a firm foundation for the compass. The eye is placed to the lens on the eyepiece. The compass is pointed at the object or point to which the azimuth is to be read. A sighting is taken through the sighting slot in the eyepiece, and the point is lined up with the sighting wire in the cover. The compass is held until the dial steadies; then the reading is taken through the lens of the eyepiece. This reading is the magnetic azimuth of the line from the observer to the point.

Reading a Back Azimuth

A back azimuth is the direction opposite the line of sight. If the azimuth is less than 180 degrees, the back azimuth is obtained by adding 180 degrees. If the azimuth is greater than 180 degrees, the back azimuth is obtained by subtracting 180 degrees. Back azimuths are used to determine a return route or to resection to determine a current position.

Circumventing Obstacles

When a scout is traveling on an azimuth and comes upon an obstacle—such as a contaminated area, minefield or swamp—the following steps (sometimes referred to as the *90-degree offset* method) is employed to go around or circumvent the obstacle and resume movement along the original azimuth (see fig. 2-3). The steps are as follows:

1. Move up to the obstacle and make a full 90-degree turn to the right (or left).

2. Walk beyond the obstacle, keeping track of the distance in paces or meters.

3. Stand at the end of the obstacle, face in the original direction of march, and follow that azimuth until the obstacle has been passed.

4. Make a 90-degree turn to the left (or right) and move the distance previously measured to return to the original line of march.

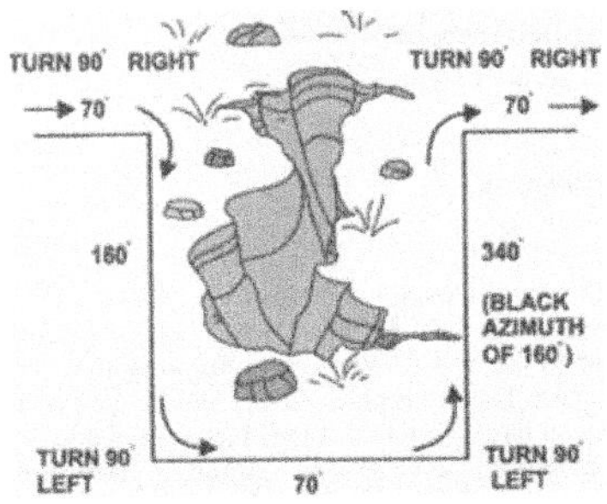

Figure 2-3. Circumventing an Obstacle.

Following an Azimuth During the Day

The eye is placed to the lens in the eyepiece and the compass moved until the desired azimuth reading is visible beneath the fixed index. Without moving the compass, the vision is shifted from the lens through the sighting slot in the eyepiece, and a sighting is taken out beyond the sighting wire in the cover. A prominent terrain feature on this line of sight is selected, the compass closed, and the landmark approached. When the scout reaches the landmark, the procedure is repeated.

Following an Azimuth at Night

It is necessary to prepare and set the compass before departing on a night movement because at night only the luminous parts of the compass can be seen. To prepare the compass for night use, the luminous parts must be fully charged by sunlight or artificial light, such as a flashlight. To set a compass—

1. Move the compass so that the desired azimuth on the dial is directly under the index line on the lower glass.

2. Rotate the upper movable glass so that the luminous line is directly above the north arrow of the dial.

3. Set the compass for marching at night on the specified azimuth.

—OR—

1. Face the general direction of movement.

2. Line up the north arrow and the luminous line on the bezel with the luminous sighting dots.

3. Hold the compass still with one hand and grip the knurled bezel ring with the other hand.

4. Turn the bezel ring the prescribed number of clicks in the proper direction, remembering that each click equals 3 degrees. For example, to set an azimuth of 21 degrees, the bezel ring would be turned seven clicks to the left.

5. Turn the whole compass until the north needle lines up with the luminous line. The compass is then set on the desired azimuth. The azimuth is the line formed by the two luminous sighting dots on the inside of the cover.

To march on a preset azimuth during night movement, open the compass and move it so the north arrow is directly below the luminous line (see fig. 2-4). Move

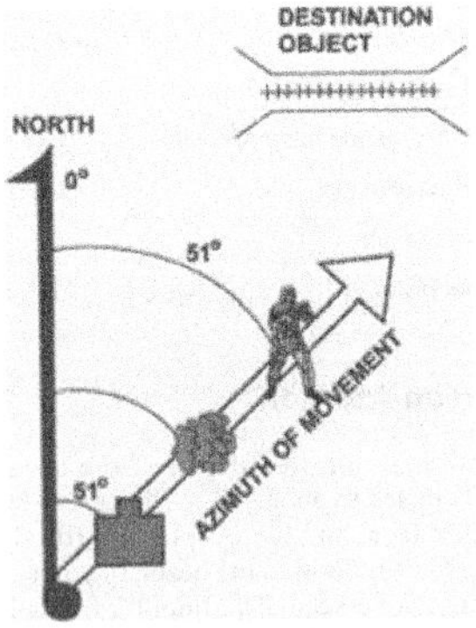

Figure 2-4. Following a Night Azimuth.

in the direction of the line formed by the two luminous sighting dots. It is necessary to refer to the compass more frequently at night than during the day. If stars are visible, find a prominent star along the azimuth of movement to use as a reference point. When the view of the sky is restricted by overcast conditions or vegetation, send a scout forward along the azimuth of movement to the limit of visibility. This scout is guided along the azimuth of movement by a stationary navigator. When the scout reaches the limit of visibility, the navigator moves to the scout's location. This process is repeated until the destination is reached.

A more rapid method for reaching the scout's destination is to equip the navigator with a compass. The navigator can set the compass as explained earlier and the scout proceeds providing security 180 degrees to the front on the specified azimuth, receiving right and left corrections from the navigator while both are on the move. The point scout must stay within visual range of the navigator. If available, a strip of white or luminous tape on the back of the point scout's helmet will assist.

Intersection

Intersection is the location of an unknown point by successively occupying at least two, preferably three known positions and sightings on the unknown point. It is used to locate features not depicted on the map or not readily identifiable. To determine an intersection, perform the following steps (see fig. 2-5):

1. Orient the map using the compass.
2. Locate and mark your position on the map.
3. Measure the magnetic azimuth to the unknown position; then convert to grid azimuth.
4. Draw a line on the map from your position on this grid azimuth.
5. Move to a second known position from which unknown point is visible. Locate this position on the map and again orient the map using the compass. The second unknown position should be a minimum of 30 degrees offset from the first position.
6. Repeat steps 4 and 5.

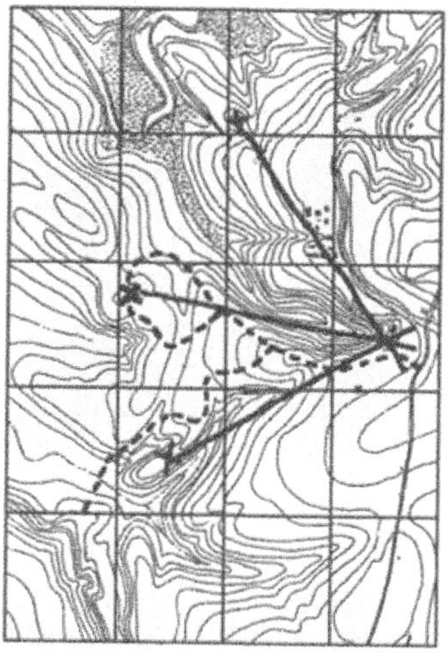

Figure 2-5. Intersection.

To check accuracy, move to a third position and repeat steps 1 thorough 4. Where the lines cross is the location of the unknown position. Using three lines, a triangle is sometimes formed—called the *triangle of error*—instead of an intersection. If the triangle is large, recheck your work to find the error. Do not assume that the position is at the center of the triangle.

Resection

Resection is the location of the user's unknown position by sighting on two or three known features that are identifiable on the map. To determine a resection, perform the following steps (see fig. 2-6):

1. Orient the map using the compass.
2. Locate two or three known positions on the ground and mark them on the map.
3. Measure the magnetic azimuth to a known position then convert to grid azimuth.

4. Change the grid azimuth to a back azimuth and draw a line on the map from the known position back toward the unknown position.

5. Repeat step 3 and step 4 to determine a second known position.

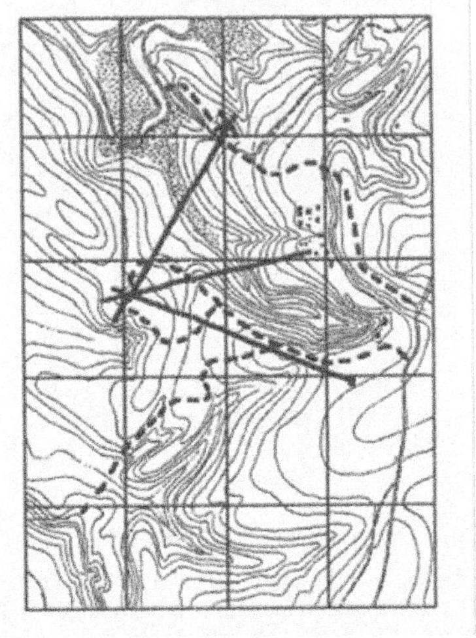

Figure 2-6. Resection.

To check accuracy, repeat the steps above for a third known position. The intersection of the lines is your location. Using three lines, a triangle of error may be formed. If the triangle is large, recheck.

2003. STARS, SUN, AND OTHER FEATURES

In rare cases when a scout is without a compass, the following examples are alternate means to determine direction. When using constellations to determine direction, identify your location's Temperate Zone. A Temperate Zone is the area between the tropics and the polar circles.

At night, the stars provide an excellent means of maintaining a line of march. In the North Temperate Zone (north of the equator), the Big Dipper constellation is one key to determining direction of true north. It is made up of seven fairly bright stars in the shape of a dipper with a long curved handle (see fig. 2-7). The two stars that form the side of the cup farthest from the handle, used as pointers, are situated in the direction of a bright star that is about five times the distance between the two stars of the dipper cup. This bright star is the North Star and is directly over the North Pole. The pointers always designate the North Star, which is the direction of true north.

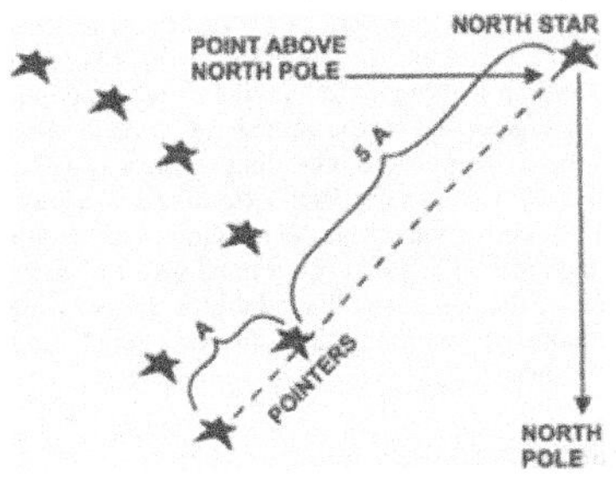

Figure 2-7. Locating the North Pole.

In the Southern Hemisphere, true south is determined in relation to the Southern Cross, a constellation composed of five stars. Two bright pointer stars in the vicinity of the Southern Cross serve as locators to help locate true south (see fig. 2-8). The outer four stars are

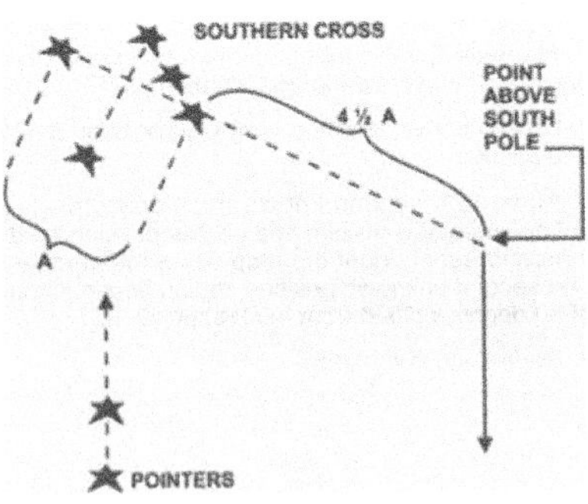

Figure 2-8. Locating the South Pole.

fairly bright and form a cross. This cross is imagined as the frame of a kite. A straight tail, four and one half times as long as the length of the kite itself, is put on the kite using finger widths for a measuring stick. The end of this tail will be close to a position directly over the South Pole. Usually, it will not be possible to see a star in the immediate vicinity, because there is no bright star visible directly above the South Pole.

During daylight hours, a watch and the sun can be used to determine direction within 8 degrees. In the North Temperate Zone, the watch is held horizontally, face up, and the hour hand pointed at the sun (see fig. 2-9). The north-south line and the direction of south can be found midway between the hour hand and the number 12, if the watch is set on standard time. If in daylight savings time, the direction of south is found midway between the hour hand and the number 1.

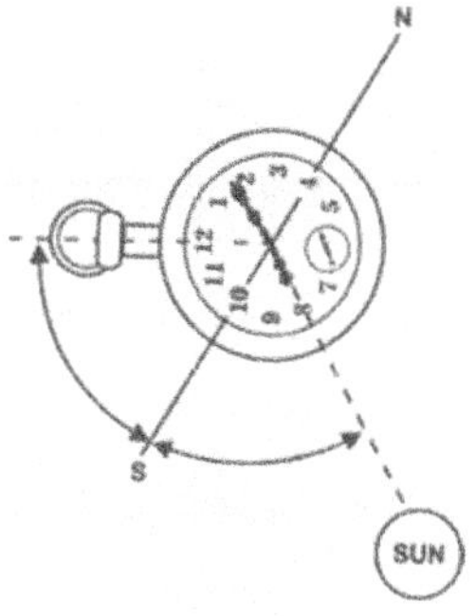

Figure 2-9. Determining Direction by Watch and Sun (North Temperate Zone).

In the South Temperate Zone, if the watch is set on standard time, the number 12 on the watch is pointed at the sun; if the watch is set on daylight savings time, the number 1 is pointed at the sun. North is midway between 12 (or 1) and the hour hand (see fig. 2-10).

When laying in a north-south line, if any doubt exists as to which end of the line is north, remember that the sun is in the east before noon and in the west in the afternoon.

In addition to the sun and stars, other methods a scout without a compass can use to determine direction include determining prevailing wind direction and

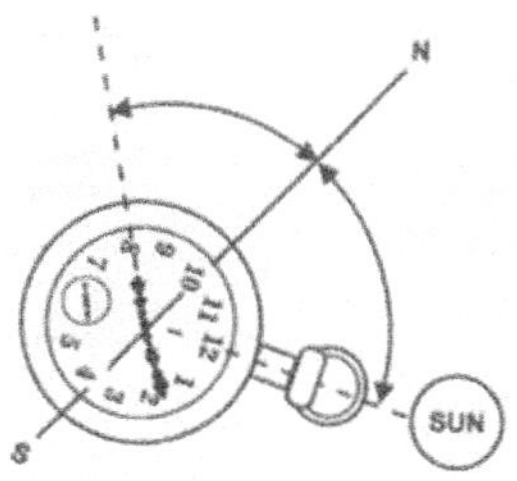

Figure 2-10. Determining Direction by Watch and Sun (South Temperate Zone).

using a mountain for orientation. By previous study of maps and photographs, a scout can keep informed of location and direction by using a distinctive edge of woods, a deep ravine or the direction of a stream's flow. A scout should constantly evaluate and memorize both the immediate terrain and general area for prominent features and landmarks.

2004. RANGE DETERMINATION

Range determination is the method of finding the distance between an observer and an enemy target or an object. By accurate range determination, the members of a given unit can set their sights correctly and place effective fire on enemy targets. The degree of accuracy is dependent on several factors, such as terrain relief, time available, and experience of the observer.

Mental Estimation

A mental distance estimate is made using a known unit of measure. Distance is estimated to the nearest 100 meters by determining the number of known units of measure between the observer's position and a target. For example, a football field, which is 100 yards, can be used as a known unit of measure for determining the distance between an observer's position and a target. For longer distances, progressive estimation may be necessary. To do this, the observer determines the number of units of measure to an intermediate point and doubles the value. The observer should consider the effects in table 2-1 in estimating distances.

Table 2-1. Effects to Consider in Mental Estimation of Distances.

Objects Appear Nearer	Objects Appear More Distant
In bright light.	In poor light or in fog.
In clear air at high altitude.	Only a small part of the object can be seen.
The background is in contrast with the color of the object.	The background is similar in color to that of the object.
The observer is looking down from a height.	The observer is looking over a depression, most of which is visible.
The observer is looking over a depression, most of which is hidden.	The observer is kneeling or sitting, especially on a hot day, when the ground is moist.
The observer is looking down a straight feature such as a road.	
The observer is looking over water, snow, or a uniform surface such as a cultivated field or desert.	

Estimating in Good Visibility

When visibility is good, distances can be estimated by using the appearance of tree trunks, branches, and foliage (as seen by the naked eye) in comparison with map data. Table 2-2 is a guide for wooded terrain. Table 2-3 is a guide for urban environments.

Estimating From a Terrain Study

The Marine should always use terrain/map analysis to assist in estimating distances. When the Marine is looking in a specific direction, the estimation of distance can be enhanced by studying the terrain and comparing it with the map. Particular emphasis should be given to color contrasts of terrain features seen along the observer-target line (OTL). For example, the distance across successive ridge lines or depressions in the distance may be identifiable by only slight changes of color to the eye. Different colors of grass might reveal a hidden terrain feature such as a stream.

Table 2-2. Estimating Distance in Wooded Terrain.

Distance in Meters	Tree Description
1,000	Trunk and main branches are visible. Foliage appears in cluster-like shape. Daylight may be seen through the foliage.
2,000	Trunk visible, main branches distinguishable, foliage appears as smooth surface. Outline of foliage of separate trees distinguishable.
3,000	Lower half of trunk visible. Branches blend with foliage. Foliage blends with adjoining trees.
4,000	Trunk and branches blend with foliage and appears as a continuous cluster, smooth in appearance. Movement of foliage due to wind cannot be deleted.
5,000 and beyond	Whole area covered by trees and appears smooth and dark.

Table 2-3. Estimating Distance in Urban Terrain.

Distance in Meters	Object Identified by the Unaided Eye
1,000	Lone tree trunk
1,500	Individuals and horsemen
3,000	Chimneys on rooftops
4,000	Windows in houses
4,000–5,000	Individual houses in populated area
8,000–9,000	Villages and individual houses
15,000–18,000	Large houses, towers, and steeples

Chapter 3. Enemy Activity

A commander often acts on information furnished by scouts. Therefore, scouts must aim at absolute accuracy in reporting enemy activity. This chapter discusses estimating enemy strengths, interpreting signs and tracks, and knowing the enemy.

3001. Estimating Enemy Strengths

If troops cannot be counted, their strength may be estimated by: noting the length of time it takes various types of moving columns to pass given point, the area required of a unit in camp or bivouac, or the front on which they are deployed. When the ground is dry, infantry on the march raise a low, thick cloud of dust, and motor vehicles or mechanized units raise a thick, rapidly moving cloud. Additionally, through practice, a scout may gain information as to the strength and composition of enemy forces by listening to noises and observing lights, fires, and smoke. A scout gains valuable experience in estimating enemy strengths by observing friendly forces in camp, on the march, and deployed. The knowledge scouts gain during field exercises of the appearance and tactical dispositions of squads, platoons, companies, and larger units will be of great assistance in estimating the strength and composition of enemy units observed under various conditions.

3002. Interpreting Signs and Tracks

In addition to estimates made through direct observation, a scout may often be able to estimate size, composition, direction, rate of movement, condition, discipline, state of training, and morale of enemy forces through signs and tracks left behind.

Signs

The examination of vacated enemy positions provides valuable information. The size of a bivouac or defense area ordinarily indicates the number of enemy occupants. Clothing, ration containers, dumps, etc., further indicate the quantity of the departed enemy force. The condition of the bivouac area and amount of material abandoned give an indication of the enemy morale, training, and discipline. A well-policed area indicates good discipline. Rubbish, ration and smoking residue, and nonessential personal items of equipment adrift indicate a lower state of morale, training, and discipline. Stores and material left behind in good condition may indicate a hasty movement or withdrawal. Burned or destroyed materials indicate a deliberate, orderly withdrawal or movement. Letters, insignia, and other articles may reveal the identity of the enemy unit.

In the case of a moving enemy, the distance between periodic halts indicates the rate of march if enemy habits relative to marches and halts are known. Condition of the halt areas indicates the state of morale, training, and discipline.

The physical condition of enemy dead and wounded and their personal equipment and weapons are reported. The general condition and state of maintenance of destroyed or abandoned vehicles should also be reported.

Tracks

A track is a mark left on the ground by the passage of a person or object. Examination of tracks reveals information about the enemy.

Troops

A few tracks overlapping each other on both sides of a road or trail may indicate a patrol in staggered formation. A large number of tracks indicates troops in column formation. A large column will wear a dry road smooth and flat. In damp terrain, a freshly made track will have sharp edges; ordinarily, signs of moisture will disappear in about 15 minutes. A runner's toes are dug into the ground; a walker's footprint is fairly even.

Vehicles

The type of track indicates whether the vehicle is wheeled or tracked. A scout acquires the necessary experience to make the proper determination by observing vehicle tracks during training.

The direction of travel can be determined by the way tracks pass across ruts, by impressions on the edges of holes in the ground, how water is splashed from puddles, or by the way grass, twigs, and branches are bent; for example—

- A vehicle (wheeled or tracked) entering a rut pushes dirt into the rut and leaves an indentation on the exit side of the rut.
- A wheel going over holes in the ground leaves a deeper impression on the edge toward the direction of travel.
- The side of a puddle with the greater splash indicates the direction of travel of the vehicle.
- When traveling cross-country, the direction in which grass is bent and/or twigs, branches, and bushes are broken indicates direction of travel.

A general rate of speed can be estimated by the amount of water or mud splattered. A fast-moving vehicle will throw larger amounts of water or mud a greater distance to the front and sides than a slow-moving vehicle, and it leaves a deeper impression on the exit edges of holes. The faster the travel, the deeper the impression.

3003. KNOWING THE ENEMY

A scout should learn as much as possible about enemy psychology, habits, organization, and tactics. The more knowledge gained about the enemy, particularly the enemy's normal security measures, the better the scout's chances are to observe and obtain accurate information with minimum risk to the mission's success. Scouts gain much of this knowledge through experience, but they also gain a great deal of their preliminary information, particularly that pertaining to enemy organization and tactics, during training and may be updated by unit commanders and intelligence officers.

CHAPTER 4. DAYLIGHT SCOUTING

A scout must be able to operate in all types of terrain and under all conditions of visibility. He must be thoroughly familiar with the principles for using cover and concealment, camouflage, individual movement, and route selection, both to and from the objective.

4001. COVER AND CONCEALMENT

Cover is protection from the fire of hostile weapons. Concealment is protection from observation or surveillance from hostile air and ground observation, but not from hostile fire. Both cover and concealment are divided into two main categories: natural and artificial. Natural cover includes small hills, ditches, rocks or vegetation. Fighting holes, bunkers, and brick walls are examples of artificial cover. Some features, such as buildings, provide both cover and concealment. In deciding whether to seek cover or concealment, a scout must make the best choice to complete the mission (see fig. 4-1).

Figure 4-1. Correct Use of Cover.

Concealment Principles

Concealment principles are as follows:

- Remain motionless while observing. Anything in motion attracts the eye.
- Use all available concealment.
- Observe from the prone position (it offers a low silhouette and makes detection by the enemy difficult.
- Expose nothing that reflects light.
- Blend with the background because contrasting colors are noticeable.
- Remain in the shade because moving shadows attract attention.
- Distort or change the regular outline of objects. Most military objects have distinctive shapes that make obvious shadows and silhouettes.
- Avoid the skyline. Figures on the skyline can be seen from great distances and are easily identified by their outlines.

Concealment Techniques

Concealment techniques are as follows:

- When observing, the scout looks around an object's side (unless it is transparent) and prepares to fire, if necessary, around the side of or, if possible, through an object.
- Looking or firing over an object can make the scout an easily visible target for the enemy. If the scout must fire over the top of concealment or cover, the outline of the head or helmet should be broken or distorted.
- Upon the approach of an airplane, the scout takes a prone position, turns face-down, and remains motionless. If surprised by an airplane, the scout remains in place and does not look up.

- The scout covers exposed body parts such as the face, back of the neck, and hands with grease paint, mud or other materials to reduce sun reflection.
- Camouflage for equipment can be improvised from garnishing or sandbags to prevent sun reflection.
- In snowy terrain, white overgarments are worn.
- The helmet cover outline should be distorted.

4002. CAMOUFLAGE

Camouflage is the use of concealment and disguise to minimize the possibility of detection and/or identification of troops, material, equipment, and installations. The purpose of camouflage is to provide concealment of military objects from enemy observation. Camouflage is also used to conceal an object by making it look like something else. A scout's mission usually requires individual and equipment camouflage. If natural camouflage is not adequate, the position is camouflaged. In using camouflage, remember that objects are identified by their form (outline), shadow, texture, and color. The principal purpose of camouflage in the field is to prevent direct observation and recognition.

Individual Camouflage

Successful individual camouflage involves the ability to recognize and take advantage of all forms of natural and artificial concealment available (vegetation, soil, debris, etc.) and knowledge of the proper use of artificial camouflage materials.

Aids to Individual Camouflage

A scout must recognize the terrain's dominant color and pattern and must change the appearance of clothing and equipment accordingly in order to blend and not contrast with the terrain (see fig. 4-2).

The helmet is camouflaged by breaking up its shape, smooth surface, and shadow. Use of a helmet cover works best. In the absence of a helmet cover, mud can be irregularly blotched on the helmet to disguise its form and dull the surface. A helmet cover may be improvised from irregularly colored cloth or burlap to blend with the background. Foliage can be draped to prevent the visor of the helmet from casting a dark shadow across the face. Foliage should not stick up like plumes because any head movement will give away the position.

A small, thin bush in the shadow of a large bush makes a good observation point. Lone trees, rocks, fence corners, and outstanding landmarks are easily picked up by the enemy as obvious observation posts.

If camouflage clothing is not available, other available clothing can be attached in irregular splotches of appropriate colors.

Exposed skin reflects light and attracts the enemy's attention. Even very dark skin will reflect light because of its natural oil. The buddy system is recom-mended when applying camouflage. Standard Marine Corps issue camouflage face paint sticks are two toned:

- Loam and light green for light-skinned troops, in all but snow regions.
- Sand and light green for dark-skinned troops.
- Loam and white for troops in snow-covered terrain.

Shiny areas (forehead, cheekbones, nose, and chin) are painted with a dark color. Shadow areas (around the eyes, under the nose, and under the chin) are painted

Figure 4-2. Avoid Contrasting Backgrounds.

with a light color. Skin that is exposed on the back of the neck and hands is painted with a two-color combination in an irregular pattern (see fig. 4-3).

When standard issue face paint sticks are not available, burnt cork, charcoal or lamp black can be used to tone down exposed areas of skin.

Mud is used only in an emergency because it changes color as it dries and may peel off, leaving the skin exposed. Since mud may contain harmful bacteria, mud should be washed off as soon as possible.

Any equipment that reflects light should be covered with a nonreflective material that aids in the concealment of the weapon (for example, black electrical tape or mud). The straight line of the rifle or other infantry weapons may be very conspicuous to an enemy observer. The barrel and hand guard should be wrapped with strips of contrasting colored cloth or

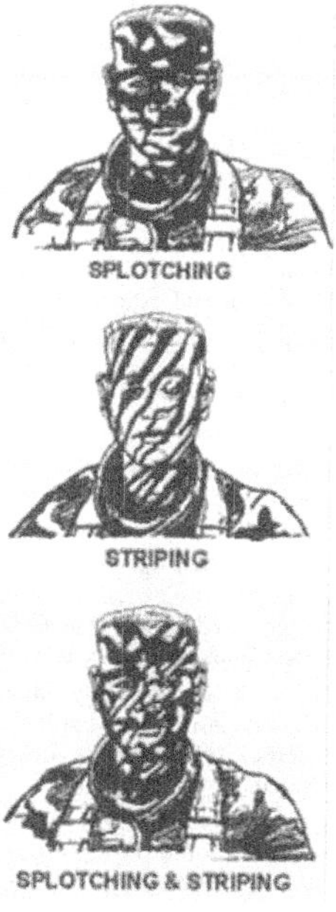

Figure 4-3. Face Camouflage.

tape to break the regular outline. Mud or dirt dulls the reflecting surface of the stock, barrel, and bayonet where coloring has been worn. Lamp black may also be used on metal parts. The function of the weapon must not impaired.

If time, material, and surroundings permit, a ghillie suit should be constructed. (Refer to MCWP 3-15.3, *Scout Sniping*.)

Aids to Camouflage a Position

To successfully camouflage a position, the scout must remember to—

- Camouflage the position as soon as it is occupied.
- Avoid using too much material for camouflage. Even though natural materials are used, too much may make the object and its shadow stand out from its surroundings, thus attracting the attention of a hostile observer.
- Inspect completed camouflage work from the enemy's point of view to check effectiveness.

Continuous Camouflage

Camouflage around and on the scout's position must be maintained in a fresh condition as wilted and dead foliage can give the position away. If the mission dictates that the position should be occupied for longer periods, wilted foliage should be replaced during periods of reduced visibility.

4003. INDIVIDUAL MOVEMENT

Principles

The principles of individual movement are as follows:

- Scouts move from one concealed position to another. When not changing positions, they remain motionless.
- The scout's head is lifted slowly but steadily, without abrupt movements, to search for a new position.
- Scouts select the next stopping place before moving and ensures it is not contained by the enemy.
- Scouts change position on the run: springs up, runs with the body bent low, zigzags, quickly drops to the ground slightly to the right or left of the objective, then rolls or crawls to the desired position.

(Remember the phrase, "I'm up—he sees me—I'm down.")

Rushing

When starting from the prone position—

- Raise the head slowly and steadily and select a new position.
- Lower the head slowly, draw arms inward, cock right leg forward, and prepare to rush.
- Use one movement to raise the body by straightening both arms.
- Spring to your feet, stepping off with the left foot.
- Bend forward as low as possible when running. Never advance directly to the next position; always zigzag.

When hitting the deck—

- Stop.
- Plant both feet in place.
- Drop quickly to the knees and slide the hand to the heel of the rifle.
- Fall forward, breaking your fall with the butt of the rifle. (To confuse the enemy, roll over after hitting the deck and roll into firing position with feet, knees, and stomach flat on the ground.)
- Keep head down if you do not intend to fire.

When rolling over—

- Hit the deck and assume the prone position.
- Bring the rifle in close to the body, placing the rifle butt in the crotch.
- Roll over swiftly to confuse any enemy observers as to final intended location. Never reappear at the same place you went down.

Low Crawl

The low crawl is used when—

- Cover and concealment are scarce.
- The enemy has good observation over the area in which the scout is moving.
- Speed is not essential.

To perform the low crawl, keep the body as flat as possible against the ground. Grasp the rifle sling at the upper sling swivel. Let the balance of the rifle rest on the forearm and let the butt of the rifle drag on the ground. Keep the muzzle off the ground.

To start forward, push arms forward and pull right leg forward. To move forward, pull with arms and push with right leg. Change the pushing leg frequently to avoid fatigue.

High Crawl

The high crawl is used when—

- Cover and/or concealment are available.
- Poor visibility reduces enemy observation
- Greater speed of movement is required.

To perform the high crawl, keep body off the ground. Rest weight on forearms and lower legs. Cradle rifle in arms, keeping the muzzle off the ground. Keep knees well behind the buttocks to stay low.

Move forward, alternately advancing right forearm and left knee; then left forearm and right knee.

Movement Aids

Aids to movement include—

- Carrying only necessities. Additional weight causes premature fatigue and impedes free movement.
- Not disturbing birds or animals whose flight would betray your presence. If birds or animals are alerted, remain motionless under cover for a few minutes, as the enemy's attention may also be attracted.
- Moving during an incident that diverts attention, such as an airplane flight, a distant disturbance or sudden bursts of fire.
- Fog, smoke, or even light haze offer concealment for movement; however, the enemy may have thermoimagery and night vision devices. Therefore, darkness and smoke cannot be used as easily.
- Following a stream or road by staying as far away from them as possible while still keeping them in sight. Keep close to the dune line when moving along a beach.
- When moving through tall grass or similar growth, move when the wind blows, changing direction frequently. A straight route will be more readily noticed.
- Whenever possible, avoid areas of soft ground so as not to leave tracks.

- When crossing a road or water obstacle, choose crossing sites where the enemy's observation is restricted (an area in shadows or near a bend) and cross rapidly.

4004. ROUTE SELECTION

Prior to Movement

A scout and the immediate commander conduct a map reconnaissance before starting on a mission. This assists them in selecting the route according to available cover and concealment and any indicated enemy activity.

Prior to and during the course of the mission, move to an observation point to visually reconnoiter the terrain for movement and select the tentative route. It may be necessary to make wide detours around open spaces or those containing enemy patrols or other enemy activity.

Carefully study the country to be traversed and pay close attention to the general features, streams, ridges to be crossed, and their relation to the general direction to be taken (see fig. 4-4).

Make notes of terrain features and landmarks along the proposed route and rely on notes for guidance (see fig. 4-5 on page 4-6). Additionally, determine the compass direction and readings for each change of direction at the start. Finally, learn the location of unit boundaries and observation/listening posts as well as general location of other friendly or scouting parties. Be sure to avoid man-made and natural obstacles as they will slow progress and overall success of the mission. If possible, use the local populous as a source of intelligence. When returning to friendly lines, avoid using the same route.

En Route

En route, the actual advance will be a series of movements from one observation point to the next. The distance and route will depend on cover and terrain. Assess the cover, terrain, and any enemy or civilian activity to determine whether or not to modify the approach or return routes. Unless the mission requires it, avoid danger areas (for example, houses, villages, potential assembly or bivouac areas, roads, and streams) that may give away your position by being observed by the enemy. When required to reconnoiter danger areas, choose a covered approach and return, and make entry or passage as quietly and quickly as possible. If part of a larger effort, the approach and return should be covered by observation and fires of the other members of the scouting party or patrol.

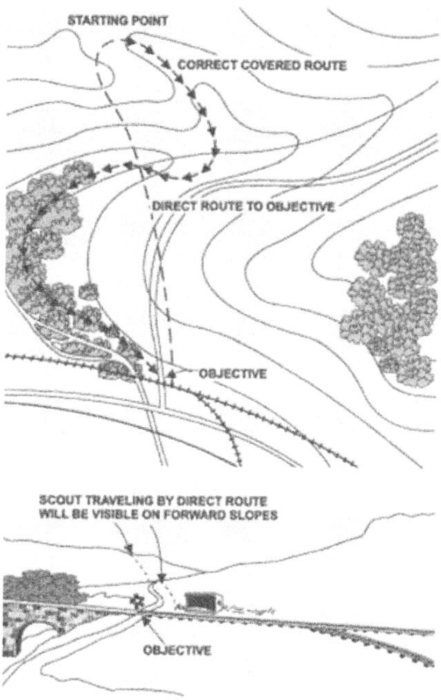

Figure 4-4. Choosing a Concealed Route of Advance from a Map.

Stream Crossings

When the crossing does not appear to be held by the enemy, advance upon it rapidly. If there are two or more scouts, one crosses while the other(s) provide protection. Note the length, width, depth, and approaches to a crossing. Observe the condition of the road or trail that crosses the stream, and report on the suitability of the crossing for use by tracked and wheeled vehicles. If the crossing is under observation by enemy, seek another crossing site or dash across to avoid detection.

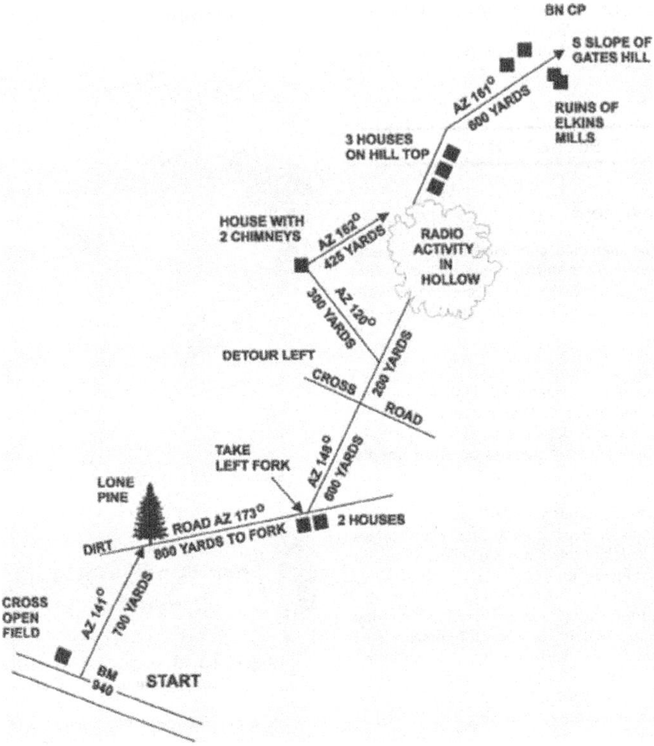

Figure 4-5. Proposed Route Sketch.

CHAPTER 5. NIGHT SCOUTING

Night scouting presents many of the same problems encountered in day operations-such as cover, concealment, movement, and camouflage-as well as additional considerations. Knowledge of human eye construction and operation will enable maximum advantage under night conditions or poor visibility.

5001. NIGHT VISION

Certain parts of the eye correspond to parts of a simple camera (see fig. 5-1). The lens focuses light entering the eye similar to a camera lens. The iris (colored part of eye) corresponds to the diaphragm of a camera, opening and closing to regulate the amount of light entering the eye through the pupil. The retina corresponds to camera film. Light rays strike the retina, form an image, and cause an impression to be transmitted to the brain through the optic nerve. In a camera, the image is formed and fixed on film.

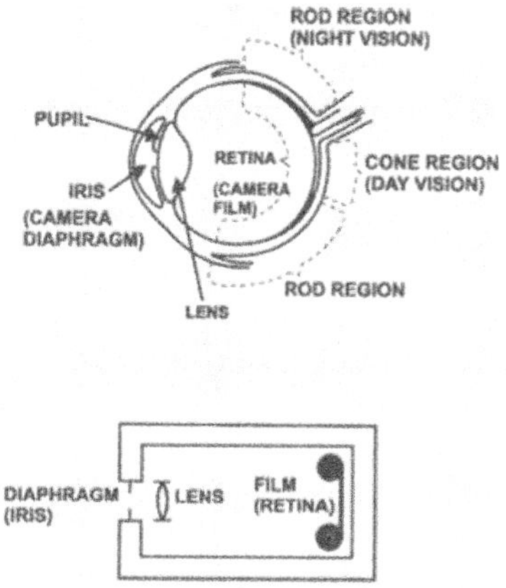

Figure 5-1. The Eye is Like a Camera.

The retina is composed of cone cells and rod cells, so-called because of their shapes. Cone cells distinguish color, shape, and sharp contrast. Because they are activated by light conditions, they are blind during periods of low illumination. Rod cells produce a chemical substance called visual purple that makes them active in darkness, low illumination or night conditions. Rod vision distinguishes black, white, shades of gray, and general outlines.

Principles

To effectively "see" at night, the principles of night vision dark adaptation, off-center vision, and scanning are applied.

Dark Adaptation

Allowing the eyes to become accustomed to low levels of illumination is called dark adaptation. It takes the rod cells about 30 minutes to produce enough visual purple to activate them and enable the eye to distinguish objects in dim light. This may also be accomplished by staying in a red-lighted area, or by wearing red goggles for 20 minutes, followed by 10 minutes in darkness (which allows the pupils to open wide). This method saves valuable time by allowing Marines to be in a lighted area to receive orders, check equipment, or perform some other function before moving into darkness.

Off-Center Vision

The technique of focusing on an object without looking directly at it is called off-center vision. When looking directly at an object, the image is formed on the cone region, which is not sensitive at night (see fig. 5-2 on page 5-2). When looking slightly to the left, right, above or below an object, the image is formed on the area of the retina containing rod cells, which are sensitive in darkness. The most sensitive area varies in individuals, but is usually found by looking 6 to 10 degrees away from an object; in effect, out of the corner of the eye (see fig. 5-3 on page 5-2).

Scanning

Off-center vision used to observe an area or an object is called scanning. When using rod vision, the visual purple in the rod cells bleaches or blacks out in 4 to 10 seconds and the object observed disappears. As the visual purple in the rod cells in one area bleaches out, the eyes must slightly shift to use fresh rod cells. Eyes

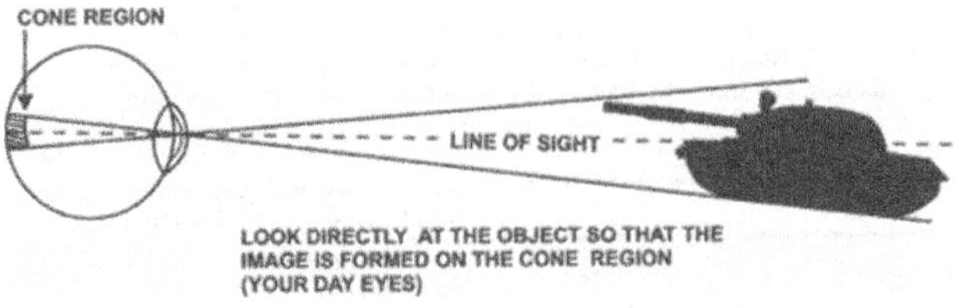

Figure 5-2. Day Vision.

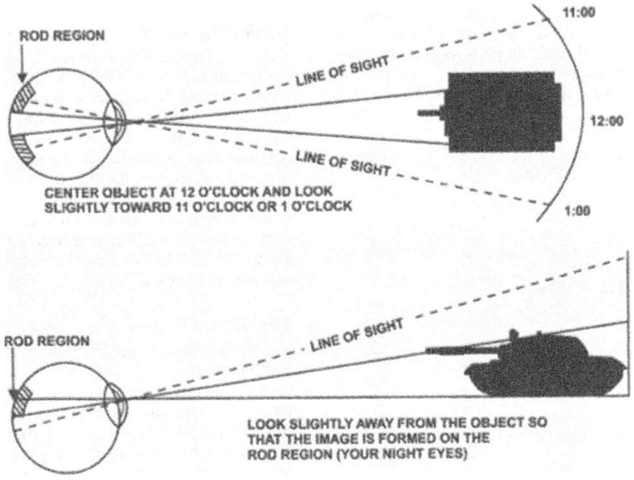

Figure 5-3. Night Vision.

should be moved in short, abrupt, irregular movements over and around the target (see fig. 5-4).

Preserving Night Vision

Night vision is quickly destroyed if bright light is allowed to enter the eye. When entering a lighted area or when observing in a temporarily lighted area (illumination, flares), one eye should be closed and covered to preserve its night vision. When the light goes off, fades or the lighted area is exited, the night vision retained by the protected eye enables it to see until the other eye adapts to the darkness. Red light helps preserve night vision, but like white light, it can be observed at long distances.

Factors that decrease night visual acuity include fatigue, lack of oxygen, long exposure to sunlight, alcohol, nicotine (within the past 48 hours), and age. When night vision has been attained, straining will not improve effectiveness; however, practice identifying objects at night will improve perception.

5002. APPEARANCE OF OBJECTS

Darkness not only makes objects difficult to see but also changes their appearance, distorts size, and blots out details. A tree visible against the night sky appears smaller than in the daytime because the twigs at the end of branches cannot be seen. A scout must train to identify objects by block outlines at night and cannot rely on details visible in daylight. Binoculars enlarge objects or parts of objects otherwise too small to be seen and help identify objects already spotted. Night observation devices increase night visibility and should be used whenever possible.

5003. SOUNDS

At night, sounds become very important. By listening, a scout gains information about the enemy and by exercising care, keeps information from the enemy. A scout stops frequently to listen. Scouts must listen for long periods in perfect silence. Hearing is amplified with the mouth open; removing the helmet will reduce sound distortion. Sounds are transmitted a greater distance in wet weather and at night than in dry weather and in the daytime. By holding the ear close to the ground sounds of people walking and vehicles moving can be heard. Sound travels approximately 370 meters a second. When a flash from a fired weapon is observed, the range to the weapons can be easily estimated by counting the time interval between the flash and hearing the report. For example, counting to three (one thousand one, one thousand two, one thousand three), indicates the distance is 1,110 meters. The cadence is determined by actual practice at known ranges.

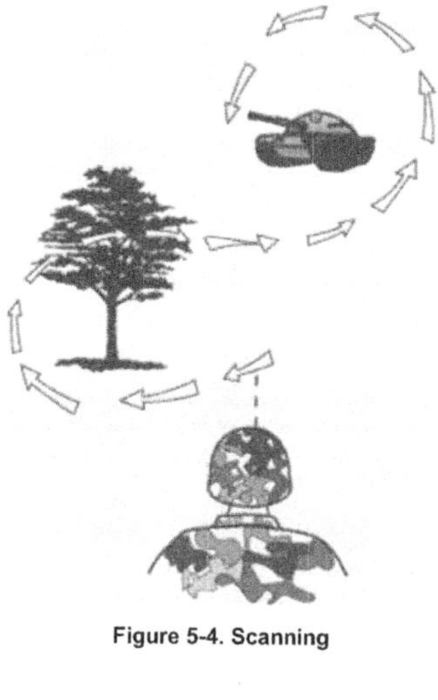

Figure 5-4. Scanning

5004. SMELLS AND TOUCH

A scout's sense of smell can warn of enemy fires, cooking, motor parks, gasoline and diesel engines, and bodies of water. A scout must feel and recognize objects in the dark, and adjust and operate equipment quietly by sense of touch.

5005. CLOTHING AND WEAPONS

All loose clothing must be secured (string or tape can be used) to prevent snagging on barbed wire, brambles, and brush. Helmet covers are worn to muffle sounds made by low branches.

The belt buckle should be turned around to the side in order to move in a prone position without scraping the buckle against stones or hard surfaces. Identification tags can be taped together to prevent rattling. Hands, face, and neck can be blackened so skin does not reflect light or appear as white spots in the darkness. (Refer to para. 4002.)

Scouts are normally armed with rifles. Rifle slings should be taped to prevent rattling. All weapons parts should be checked for glare elimination measures.

5006. CONCEALMENT

Although total darkness provides concealment, scouts must observe the same principles of concealment during moonlight conditions as in the daytime. Scouts should assume enemy employment of night observation devices and observe the principles of night movement such that presence will not be disclosed by noise when close to the enemy.

5007. AIDS TO NIGHT SCOUTING

Aids to night scouting include the following:

- Carry out scouting missions close to or within hostile positions on dark or rainy nights.
- Stifle a threatening sneeze by pressing fingers upward against the nostrils.
- Stifle a threatening cough by applying slight pressure with the finger on the Adam's apple.
- Stop a ringing sound that interferes with hearing by yawning.
- Speak softly rather than whispering if voice communication is necessary.
- Move boldly and rapidly when taking advantage of any sound-such as shelling, rustling wind or distant firing-to push forward if firing is taking place.
- Avoid shell craters and depressions in damp weather conditions if the enemy has employed chemical munitions.
- Move the eyes constantly; concentrating on one object too long will strain them.
- Take notice of the enemy's use of flares. When the enemy employs flares, few enemy patrols are apt to be out; when flares are not employed, the enemy's patrols are likely to be numerous.
- Drop to a prone position upon hearing a flare being fired and before it illuminates. Remain motionless while it is burning. If open or moving when a flare bursts in the air, freeze or drop quickly in the split second after the flare illuminates while the enemy is blinded. You are an easy target for the enemy if the flare bursts in the air or on the ground behind you. Never look at a flare. If you activate a trip flare, drop to the ground and crawl away from the illuminated area.
- Consider all patrols or individuals encountered as hostile until proven friendly. If encountering someone, crouch low, silhouetting the approaching person against the sky. At the same time, make yourself an indistinct target in case the person encountered is an enemy.
- Return fire only to avoid capture if fired on when close to enemy positions.

5008. AIDS TO NIGHT MOVEMENT

- Aids to night movement include the following:
- Move silently.
- Advance in stealthy legs. Each leg should follow some terrain feature that serves as a guide. When there are no terrain features to serve as guides, move in a straight or nearly straight line from one defined point to another, or maintain direction by using a compass.
- Avoid running, except in an emergency.
- Take advantage of sounds that may distract the enemy.
- Fall silently without making an outcry.

Walking

When walking at night—

- Place the heel down first. Balance the weight of the body on the rear foot until a secure spot is found.
- Lift the forward foot high to clear any stiff grass, brush, or other obstruction.
- Continue to balance body weight on the rear foot, lower the forward foot gently, toe first, to explore the ground for objects that might make noise. Step over fallen logs and branches, not on them.
- Lower the heel of the forward foot slowly; gradually transferring body weight to that foot.

Creeping

The low crawl and high crawl are not suitable at night when very near the enemy because an easily heard shuffling noise results. Creeping is the recommended method of movement:

- Creep at night on the hands and knees.
- Use your hands to feel for twigs, leaves or other substances that might make a noise. Clear a spot to place your knee. Keeping your hand at that spot,

bring your knee forward until it meets your hand. Then place your knee on the ground and repeat the action with the other hand and knee.

- Lay the rifle on the ground at your side and clear an area for it. Lift the rifle up and move it forward. Movement is slow and tedious, since it must be done silently.

Hitting the Deck at Night (Right-Handed Shooter)

To safely hit the deck at night from the standing position—

- Advance your left leg, place the butt of the rifle in your right armpit with the hand remaining on the pistol grip, and grasp it with the right hand at the balance.
- Quietly drop down on the right knee and left hand.

- Move the left leg carefully to the rear, and then move the right leg to the rear.

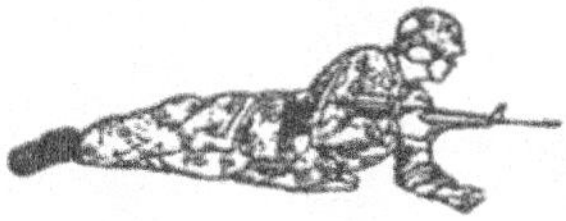

- Lie flat on the ground, or take up a firing position if necessary.

Wire Obstacles

A mission often requires a scout to pass through and work behind enemy positions. To accomplish this, the scout must be able to quietly pass through enemy wire obstacles and cross trenches. Cutting a gap in wire is time-consuming. If possible, walk over the low bands of enemy wire and crawl under the high bands (see fig. 5-5). Avoid movement along wire barriers, as enemy

Figure 5-5. Crossing Wire Silently at Night.

covering fires are generally planned parallel to them to take advantage of canalization and enfilade fire.

To step over low wire at night, crouch low to view the strands against the sky. Grasp the top strand with one hand; with the other hand, reach forward and feel for a clear spot for foot placement without stepping on other strands or any object apt to make a noise. Raise the body up, still grasping the top strand of wire. To avoid catching the foot in another strand, lift the foot up and over, passing it close to the hand grasping the wire.

If a high wire obstacle is encountered at night and wire cutters are not available, pass under the wire with your back on the ground. Grasp the lowest strands in your hands and hold them clear of the body while you slide under them.

When cutting wire and working solo, cut a wire near a post (see fig. 5-6), then dispose of all but one loose end. Grasp the wire close to a post and cut between your hand and the post, muffling the sound and keeping the loose wire in your grasp. When cutting wire in tandem, one firmly holds the wire with the hands positioned close to the cutters, in order to muffle the sound and prevent the loose ends from flying back, while the other one cuts. In both instances, the loose ends of the wire are bent back to form a passage.

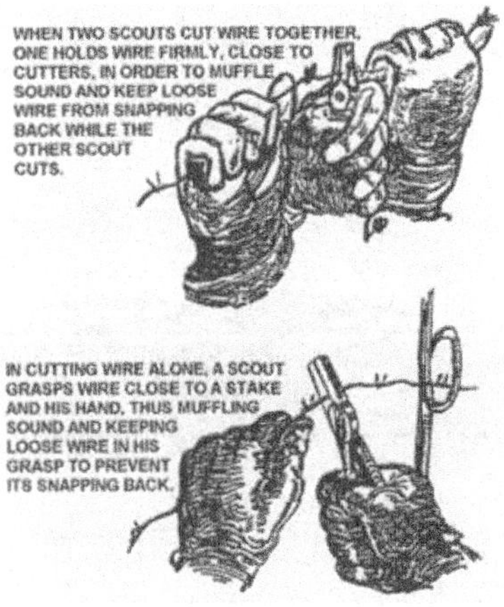

Figure 5-6. Cutting Wire Silently at Night.

Wrap a sandbag around the wire cutters and wire to deaden the sound.

Do not cut a complete gap in the wire; cut only the bottom wire(s). Leave the top wire(s) intact to lessen the chance of discovery by the enemy.

Crossing Trenches

Before approaching a trench, wait outside the trench for awhile and listen. Do not enter or cross a trench near its junction with a communication trench. Crawl silently up to the edge of the trench and look into it. Remove all loose dirt and rocks from the edge. If it is a narrow trench, spring up and jump across, sinking quietly to the ground on the other side and remaining there a moment to listen before proceeding. If the trench is wide, climb silently and slowly down into it and out the other side, using the revetment for support (see fig. 5-7). Do not enter enemy trenches unless it is absolutely necessary in order to accomplish the mission. Ordinarily, work is better accomplished from outside the trench. Sentries usually pay more attention to sounds in front of them; therefore, if it is necessary to enter a trench, cross it first at the place where enemy observation is restricted, then approach from the rear.

5009. LOCATING AND PLOTTING THE ENEMY AT NIGHT

For night work, a scout must understand the use of a lensatic compass. (Refer to para. 2002.) Using the lensatic compass, the scout can guide the platoon into position, locate adjoining elements of the command, keep direction when on patrol, determine the location of gaps in the enemy wire and the position of enemy out guards.

Locating Gaps in Enemy Wire

When searching for gaps in enemy wire, at least two lensatic compasses are needed: one to register the gap in the wire and the other for navigation. When a gap in enemy wire is located, lie outside the gap, keeping a distance of 10 meters from the barbed wire. Sight with the lensatic compass on a prominent point on the skyline in line with the gap. In selecting the prominent point in the skyline, pick one that appears on the map (i.e., hill mass, house, road junction). If the only prominent point available is one not identifiable on a map (i.e., a tree, destroyed vehicle, enemy position),

Figure 5-7. Crossing Trenches Silently at Night.

use it. The exact location of this point can be fixed the next day by visual reconnaissance of the area from an observation point. When the needle rests—

- Clamp it in place by lowering the eyepiece to the closed position.
- Rotate the movable bezel ring until the luminous line is directly over the north end of the needle. The azimuth of the gap from the prominent point is now registered.

The compass is carried back without further adjustment. The azimuth setting can be recorded later on a map (see fig. 5-8).

Locating Enemy Out Guards

At night, locate enemy guards by their sounds and failure to maintain light discipline. As sounds of the enemy are heard, and/or observations of the enemy made, shoot an azimuth with one compass. When the needle comes to rest, clamp the compass by lowering the eyepiece to the closed position.

Figure 5-8. Locating a Gap in Enemy Wire at Night.

Note the time and nature of each sound, the estimated distance, and which compass was used to fix the location. Plot this data on a map then wait until the debrief to turn in the notes and compasses. See figure 5-9.

Figure 5-9. Locating Enemy Outguards at Night.

50010. ROUTES OF MOVEMENT

Prior to beginning a night missions, a scout studies the ground in detail from an observation point, air photos, and a map during daylight. The route of advance should be below the skyline. Avoid becoming a silhouette (see fig. 5-10).

Unless the moon is bright, avoid, if possible, passing through woods, ditches, ravines, and brush, because noises of movement may lead to discovery. If the enemy is known to have night observation device capability, avoiding these kinds of terrain may not be possible. To avoid enemy ambushes, return by a different route from the advance and change routes on successive nights.

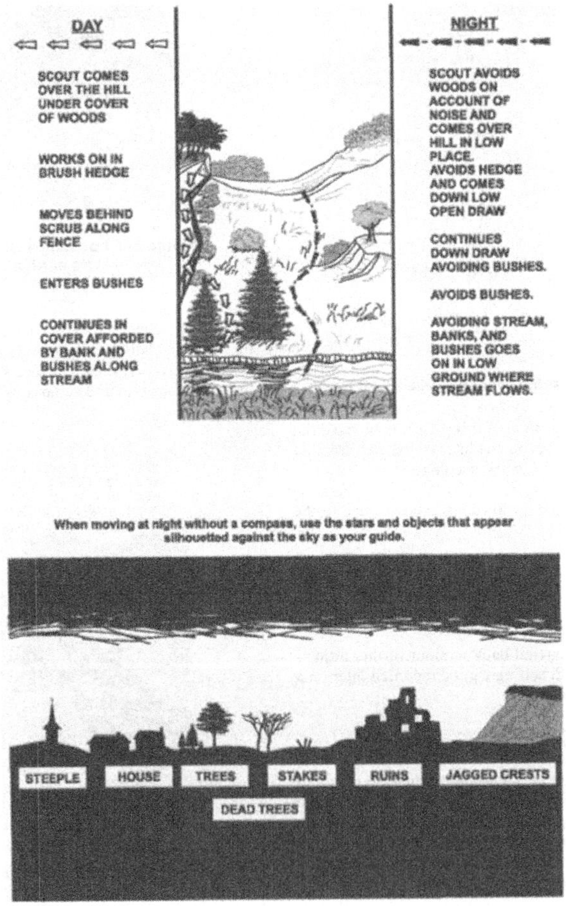

Figure 5-10. Difference Between Correct Routes by Day or Night.

CHAPTER 6. OBSERVING AND REPORTING

When sent out on a mission, a scout's duties are to observe and report. the first five chapters covered scout protection measures and movement in enemy territory. This chapter discusses performing the assigned mission and reporting the mission after its completion.

6001. OBSERVATION POSTS

Positions

The following guidelines apply to observation posts. See figure 6-1.

- When selecting observation posts, scouts should chose the least prominent position.
- Scouts may occupy one or more observation posts.
- An observation post should not be manned for more than 24 hours.
- A selected observation post should be observed for 10 to 15 minutes to ensure it is not occupied.
- Scouts move to the chosen observation post by a concealed route.
- If the post is located on a hill, crawl to a position where the skyline is broken.
- If a tree is used, the position should have a background so as not to be silhouetted against the sky while climbing or observing.
- When leaving the observation post, a different route from that of the approach should be used.
- If a radio is used, its antenna should be located to provide clear communication to the controlling commander but masked from enemy observation and direction-finding equipment. Upon departure, scouts should remove the antennae from the observation post so as not to give away the position.

Observing

Using all senses available, be particularly alert for movement, objects, sounds, and smells inappropriate to the surroundings. While observing, avoid all unnecessary movement. If observing from a building, keep back from doors and windows.

In daylight, look first at the ground nearest you. Begin observing close to your post and search a narrow strip 50 meters or less deep, going from right to left parallel to your front; then search from left to right a second and similar strip farther away but overlapping the first. Continue to observe until the entire field of view has been searched (see fig. 6-2 on page 6-2).

At night, use a night observation device. If one is not available, search the horizon with short, jerky movements, and short pauses. Look a little to one side of an object and then to the other. Lower the head close to the ground to view the object more clearly. Use low-powered field glasses to increase sight range.

Figure 6-1. Method of Approaching an Observation Post.

Figure 6-2. Method of Searching Ground.

6002. REPORTING

It is imperative that the scout accurately and completely report who, where, when, and what was observed to the proper authority upon mission completion. In most cases, a scout will not be equipped with a radio.

Verbal Reports

Verbal reports should be made when writing is impractical, when the information is not complicated or when the enemy is likely to intercept a messenger carrying a written message.

Written Messages

Written messages, preferred to verbal reports, are recorded in message book blanks issued for that purpose (see fig. 6-3) and delivered to a higher authority as soon as possible. The NATO spot report (SPOTREP) should be reviewed as the written message is a NATO format.

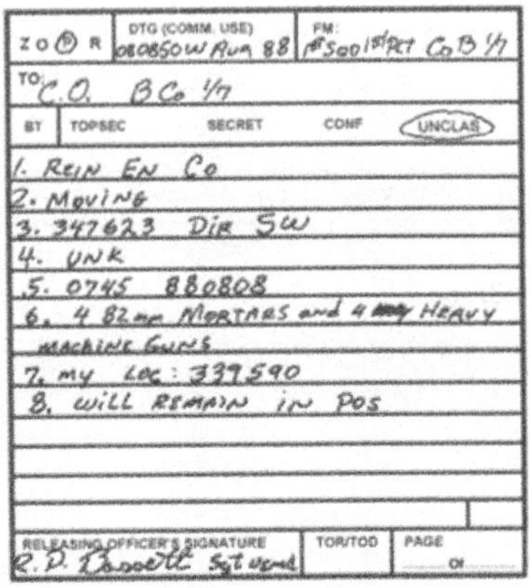

Figure 6-3. Sample Field Message.

The message body is brief, accurate, and clear; facts and opinions are distinguished. If secondhand information is reported, its source is included. Reports include all information of value, first about the enemy, and then about the originator. Information about the enemy should cover—

- Size and/or strength.
- Actions or activity.
- Location and direction of movement.
- Unit identification. (The designation of the enemy unit may be derived from unit markings, uniforms worn or through prisoner interrogation.)
- Time of observation.
- Equipment and weapons.

Messages are printed in block letters. Individual items of information are numbered and separated into paragraphs. If doubt exists as to message receipt by the commander, a summary of its contents is included in the next message. Information about the originator or writer should cover—

- Location at the time of enemy observation (reference to an important terrain feature, by map coordinates, by the back azimuth from each of two definitely located points, or the back azimuth and distance from one known point).
- Intentions. (Remain in position? Continue on the mission? Take other action?)

The message is carefully reread and if possible read by another person to ensure understandability. If a messenger is used, the messenger must read and understand the message in order to answer any questions the commander might have.

Sketch

Information difficult to describe may be given accurately on a simple sketch. The sketch may give all the necessary information or it may be used to supplement a written message. A military sketch is generally one of two types: simple or panoramic. The simple sketch is easily made and read.

Figure 6-4 shows a simple sketch that has been included in the message itself. The sketch may be on a separate sheet of paper, but all of the necessary information must be contained in either the sketch, the message or both.

The panoramic sketch is a picture of the terrain's elevation in perspective, as seen from one point of observation. Although a panoramic sketch is not difficult to create, skill and training are necessary to enhance usefulness. Figure 6-5 on page 6-4 illustrates how to make a panoramic sketch.

Overlay

The same information sent back to higher headquarters on the sketch may be sent on an overlay, if the sender and the person to whom the message is to be sent have copies of the same map. Figure 6-6 on page 6-6 illustrates a simple overlay. The overlay is drawn on transparent paper as follows:

- Orient the map and place it on a hard, flat surface.
- Place the transparent paper over the part of the map of the object or information to be transmitted and hold the paper in this position.
- Orient the overlay to the map by tracing in the intersecting grid lines at two opposite corners of the overlay. Write the correct number designation on the overlay. The cross made by the intersection is called a *tick mark* and enables the receiver to locate the exact area on the map covered by the overlay.
- Sketch the object seen or the information to be transmitted on the tracing paper (the sheet on top of the map) in the exact location it would appear on the map (the sheet underneath the tracing paper).

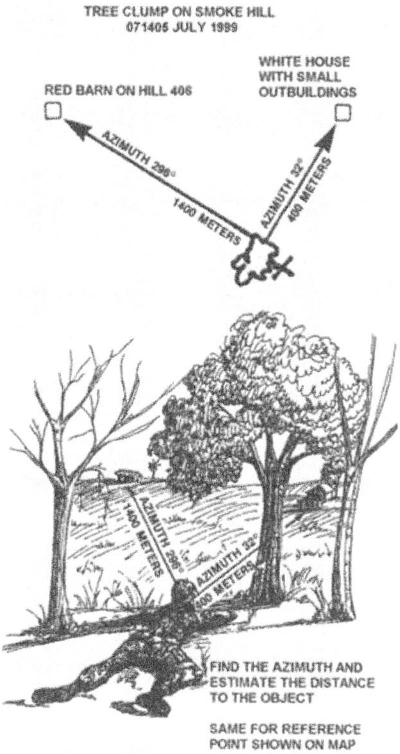

Figure 6-4. Making a Simple Sketch.

Explanatory notes are annotated in the overlay's margin, arrows point to the objectives.

- Indicate with an "X" and an appropriate explanation the position from which the observer saw the object or obtained information.
- Include title and scale of the map from which the overlay was made, date and hour the information was obtained, and signature of the observer on the overlay in the lower right-hand corner.

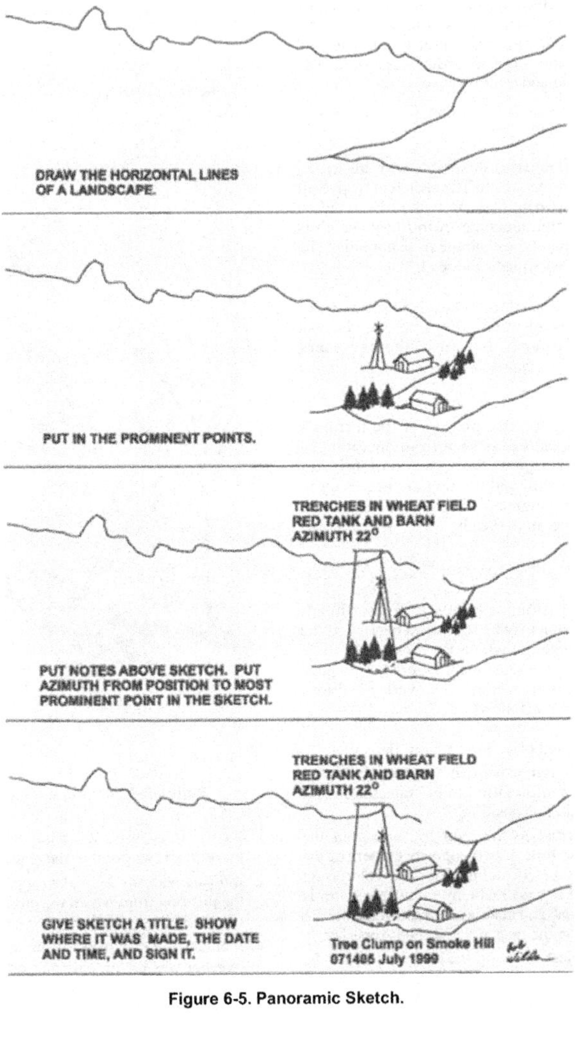

Figure 6-5. Panoramic Sketch.

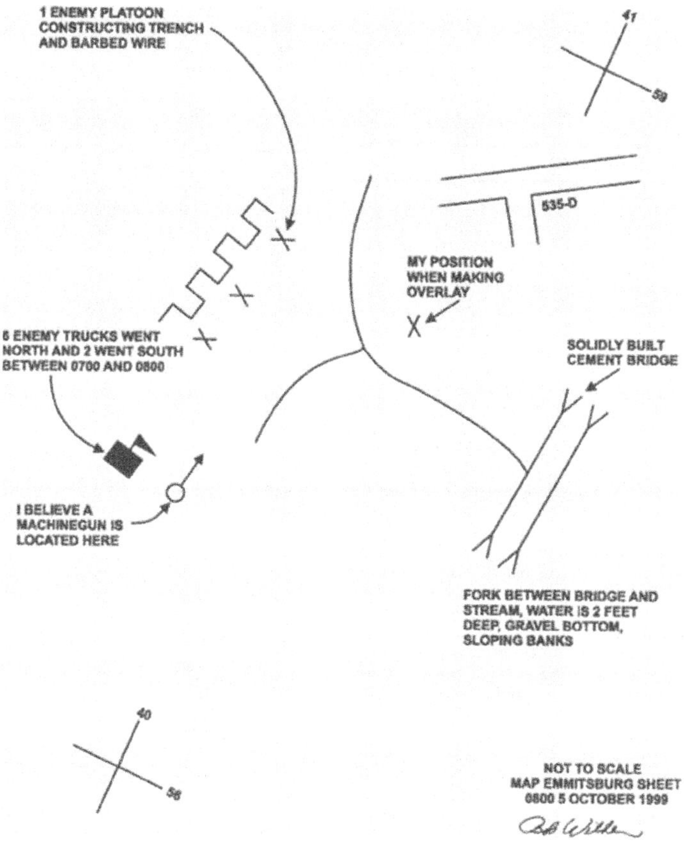

Figure 6-6. Simple Overlay.

CHAPTER 7. SCOUTING FIRE TEAMS

Upon completion of individual training, a Marine adapts this training to the functioning of the unit. The basic tactical unit for scouting is the fire team. Within the fire team, scouts normally work in pairs to furnish security and gather information. Information gathered by the individual scouts is passed to the fire team leader.

7001. POSITIONING

When a rifle platoon in the approach march is not preceded by adjacent units (left, right, front, and rear), it employs its own scouting elements. The scouting element for a platoon is usually one fire team; however, an entire squad may be used.

A fire team used as a scouting element is called a scouting fire team and is controlled by the platoon commander, assisted by the squad leader. A squad leader whose squad is providing the scouting fire teams normally marches near the platoon commander to assist in the control of the scouting fire teams.

A scouting fire team moves aggressively to cover the front of the advancing platoon and to locate the enemy's position(s). It generally moves in a wedge or skirmisher's formation. Normally, a scouting fire team is deployed on a frontage of 50 to 75 meters (10 to 17 meters between each individual scout). The entire squad may be employed to cover a wider frontage. The platoon commander coordinates the movement of the scouting fire team(s) so as to protect the main body of the platoon from enemy fire from points within 400 to 600 meters away, or in close terrain from points within the limits of enemy observation.

Scouting fire team(s) should have enough firepower to overcome resistance from small enemy advanced posts and patrols; the intent is to make enemy riflemen and machine gunners open fire and disclose their positions. Without scouts in advance, the platoon may move into areas where enemy fire may prevent further advance or maneuver and inflict heavy casualties (see fig. 7-1).

Scouting fire teams are covered by the platoon or, when the platoon is masked, the fire team leader ensures individual scouts are maneuvered and coordinated so that the fire team covers its own advance. The fire team leader constantly watches for signals from the platoon commander and remains in visual contact at all times.

The distance between the scouting fire team and the platoon is terrain-dependent. The scouting fire team should not be beyond visibility of the platoon. In open terrain, the platoon commander usually directs the scouting fire team to move by bounds along a succession of locations designated by the platoon commander as intermediate objectives.

Individual scouts should advance as stealthily as possible, while remaining consistent with their mission of reconnaissance to the front, taking advantage of cover without delaying the advance. An occasional glimpse of scouts constantly advancing over a wide front can make the enemy uneasy. It is this activity, and not the target the scouts offer, that may cause the enemy to open fire and disclose its location.

When fired upon, scouts must drop to cover and return fire only when necessary to complete the mission. If

Figure 7-1. Position of Scouts Preceding an Attacking Platoon.

necessary, one scout reports back to the fire team leader who informs the platoon commander.

7002. LOCATING ENEMY POSITIONS

The key terrain of defensive positions are those points that afford extended observation over the ground where the attack must advance. The enemy will place machine guns and infantry to defend critical points.

Members of a scouting fire team preceding an attacking platoon identify the probable enemy infantry and machine gun positions (see fig. 7-2). They use concealment and cover to conduct their advance in order to discover the exact location of enemy positions.

7003. ACTION WITH AN ATTACKING PLATOON

The scouting fire team reconnoiters to the front of the advancing platoon. As soon as the scouting fire team leader indicates the area is secure, the platoon advances and the scouting fire team moves forward. Squads within the platoon advance by bounds; at least one squad is positioned to support the other(s) by fire. Successive positions along the line of advance are selected and designated by the platoon commander as intermediate objectives, and reconnoitered by the scouting fire team before occupation. By conducting proper reconnaissance, surprise by the enemy or movement in the wrong direction may be prevented.

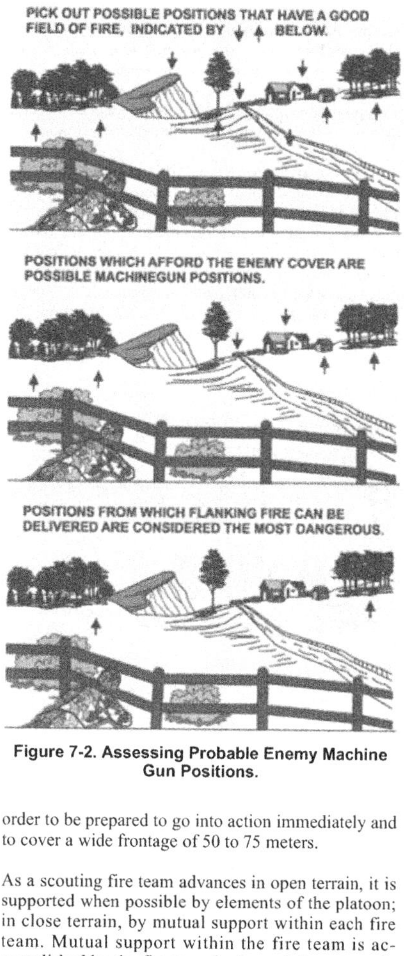

Figure 7-2. Assessing Probable Enemy Machine Gun Positions.

Movement

The distance between the scouting fire team and the front of the main body of the platoon is dependent on the mission, enemy, terrain and weather, troops and support available, and time available (METT-T). In close terrain, such as dense woods, the scouting fire team's movements closely resemble those used for night operations. In approaching houses, woods, and villages, one scout of each pair covers the other while the latter reconnoiters (see fig. 7-3).

A scouting fire team moves forward aggressively to cover the front of the advancing platoon, usually adopting either skirmisher's or wedge formation in order to be prepared to go into action immediately and to cover a wide frontage of 50 to 75 meters.

As a scouting fire team advances in open terrain, it is supported when possible by elements of the platoon; in close terrain, by mutual support within each fire team. Mutual support within the fire team is accomplished by the fire team leader and the automatic rifleman forming a team that supports by fire the advance of the rifleman and assistant automatic rifleman until they reach a location designated by the fire team leader. The rifleman and assistant automatic rifleman then support by fire the movement of the team leader and automatic rifleman. These successive points to which the sub elements of the fire team move are normally designated as fire team intermediate objectives by the fire team leader. The fire team leader sets as many fire team intermediate objectives as necessary to maintain mutual support within the team. This process is repeated until the team can be covered by other elements of the rifle platoon.

A scouting fire team takes advantage of available cover and concealment without delaying its advance. The orders of the platoon commander govern the distance at which it precedes the platoon. The terrain and the probable position of the enemy affect the scouting fire team's distance in front of the platoon. It may be as much as 400 to 600 meters in advance of the platoon. In open terrain, the platoon commander usually directs that the scouting fire team move by bounds to a succession of intermediate objectives. In close terrain or conditions of limited visibility, the scouting fire team is normally ordered to precede the platoon at the limit of visibility, maintaining visual contact with the platoon commander.

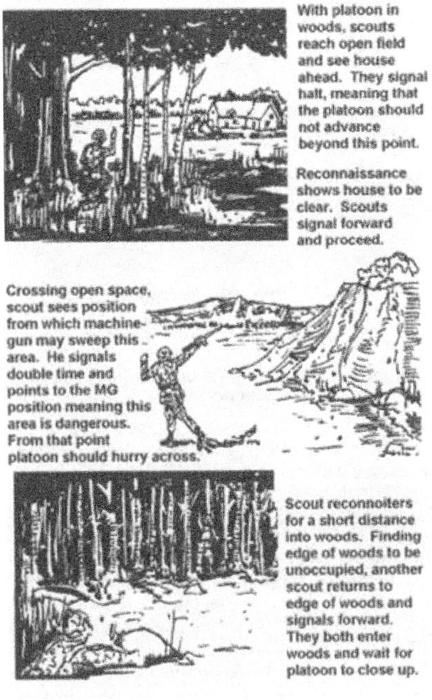

Figure 7-3. Conduct of Scouts Preceding a Platoon.

Action in Woods

When a scouting fire team is directed to advance over open ground to the edge of a woodline, two members of the team, preferably the rifleman and assistant automatic rifleman, reconnoiter inside the woodline while the remainder of the fire team covers them. It is not recommended they separate until finished with their reconnaissance of the far side of the danger area. Both members staying together can cover the same area using a zigzag reconnaissance and they are better equipped to overcome any opposition.

In heavy underbrush and/or poor visibility, the rifleman and assistant automatic rifleman proceed into the woodline together for 50 to 60 meters. The two then separate, searching out either flank to the first high ground or limits of observation, probably 50 to 100 meters (see fig. 7-4 on page 7-4). After the initial search and out posting the limit of advance, the remaining scout signals the fire team forward.

In light underbrush and/or good visibility, the assistant automatic rifleman remains at the edge of the woodline while the rifleman searches the woodline. (see fig. 7-5 on page 7-5). The rifleman searches the woodline in a zigzag pattern, reports back to the assistant automatic rifleman, then moves to an outpost position at the limit of advance. In turn, the fire team leader signals the platoon commander that it is safe for the platoon to move forward.

The fire team leader then moves the remainder of the fire team into the woods, joining up with the forward scout manning the outpost. The scouting fire team occupies and holds a line 50 to 75 meters within the woods and observes toward the direction of movement until the platoon closes up. The scouting fire team leader awaits further word from the platoon commander before moving the team further into the woods.

When directed, the scouting fire team leader moves the team forward until they reach the far edge of the woods. The team is held at the edge of the woods and the fire team leader notifies the platoon commander of the situation. The platoon commander moves the platoon to a position where it can cover the scouting

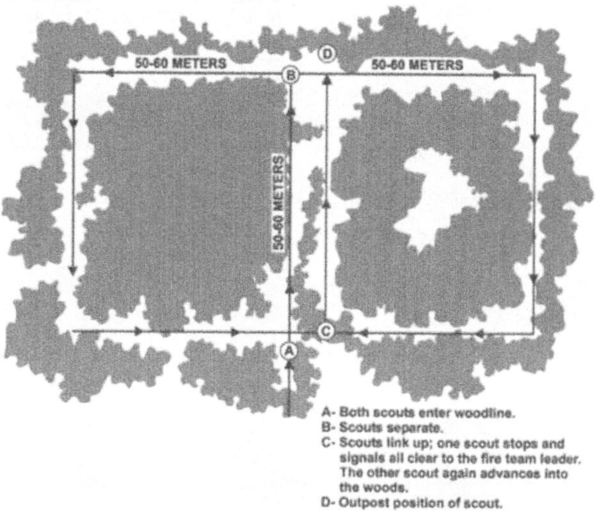

Figure 7-4. Searching Edge of a Woodline (Dense Underbrush/Poor Visibility).

fire team as it exits the woods and directs the team leader to move out and continue the scouting mission.

A scouting fire team passing through woods ahead of its unit maintains a distance allowing visual and oral communications. If an obstacle is encountered, reconnaissance to its front and flanks must be carried out. When advancing along a road or path, scouts precede the platoon to provide necessary protection and to prevent surprise fire on the platoon. When crossing a road or path, they reconnoiter well to the flanks before signaling "all clear" to the platoon.

The scouting fire team will not exit the woods until the arrival of the platoon commander, who will then be given an opportunity to alter the disposition or direction of march. The point where the platoon exits the woods is considered a danger area where the platoon is vulnerable to enemy fire. The scouting fire team is sent ahead to reconnoiter the danger area, as well as the next area to be occupied by the platoon. They signal back whether conditions require a halt, an advance or a quick rush across the open area. The scouting fire team leader must be continually on the lookout for signals from the rear.

Action Under Fire

When a scouting fire team is fired upon, they immediately take cover, locate targets, and return fire. The scouting fire team leader then determines—

- Location of enemy (range and reference points).
- Extent of position (location of flanks).
- Types of positions (obstacles, bunkders, fighting holes, etc.).
- Number of enemy.
- Enemy weapons (machine guns, mortars, tanks, etc.).

The platoon commander assesses the situation as quickly as possible based on the limited information obtained. Usually the platoon commander brings up the remaining squads, sets up a base of fire, and assaults the enemy position. Should the enemy position prove too strong for the platoon, the platoon remains engaged with the enemy as a base of fire until the remainder of the company is committed to clear the enemy resistance.

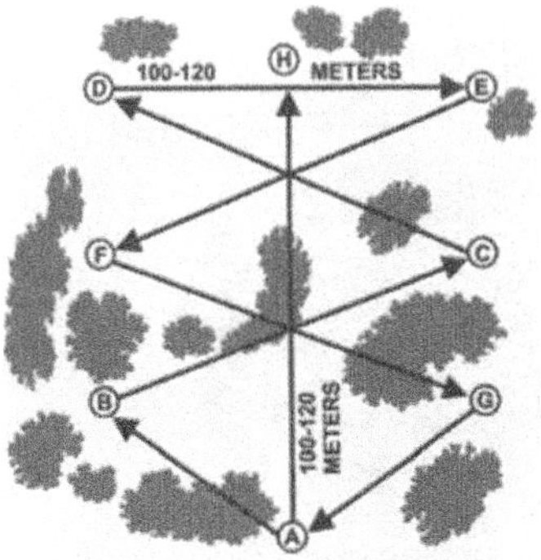

A- Both scouts enter woods. Assistant automatic rifleman remains here and covers forward movement of rifleman.

B through G - Rifleman moves to these positions, stopping and observing before moving to next position.

G to A - Rifleman reports results of his search to the assistant automatic rifleman.

A to H - Rifleman moves to out-post position.

Figure 7-5. Searching Edge of a Woodline (Light Underbrush/Good Visibility).

7004. ACTION WITH AN ENVELOPING UNIT

When a platoon is given the mission to envelop an enemy position, a scouting fire team is employed for protection and reconnaissance in the same way as when the platoon is advancing in the approach march.

PART 2. INFANTRY PATROLLING

CHAPTER 8. FUNDAMENTALS OF INFANTRY PATROLLING

This chapter begins Part 2 and provides basic information about infantry patrols; specifically, their purpose, types, and missions. Infantry patrol training is also addressed and keys to successful patrolling are presented. Subsequent chapters of this part cover patrol organization, preparation, movement, and reconnaissance actions.

8001. DEFINITIONS

A patrol is a detachment of ground, sea or air forces sent out for the purpose of gathering information or carrying out a destructive, harassing, mopping-up or security mission (Joint Publication [JP] 1-02). The mission to conduct a patrol may be given to a fire team, squad, platoon or company.

8002. RELATION OF PATROLLING TO SCOUTING

Each patrol member must be knowledgeable in the principles of scouting and maintain membership of a larger team. To develop the teamwork skills required among the members of a patrol, additional training beyond the basic principles is necessary to become a well-trained scout. A patrol member must respond quickly to the decisions and orders of the patrol leader. There must be complete confidence among all members of the patrol and the confidence that they, as a team, will be successful in their mission.

8003. PURPOSE

A commander must have current information about the enemy and the terrain in order to employ the unit effectively. Patrols are an important means of gaining this information and are used to destroy enemy installations, capture enemy personnel, perform security missions or prevent the enemy from gaining information. Modern warfare places a high premium on effective patrolling because units have larger areas of operations and can be threatened from all directions. As distances between units increase, more patrolling becomes necessary to prevent infiltration by guerrillas or small enemy units, as well as to maintain contact with friendly adjacent units. Active patrolling by numerous small groups is needed to locate the enemy and gather information on the enemy's disposition, strength, morale, and weapons, as well as gather and confirm information about the terrain.

8004. TYPES OF PATROLS

Classification as to Mission

Reconnaissance Patrol

Reconnaissance patrols gather information about the enemy, terrain or resources. Relying on stealth rather than combat strength, they gather this information and fight only when necessary to complete the mission or to defend themselves. The distance covered by reconnaissance patrols varies based on the terrain and mission. The squad is ideally suited for reconnaissance patrol missions because of its relative small size and its experience of working together.

Combat Patrol

A combat patrol is a fighting patrol assigned missions that require engagement with the enemy in combat. Larger and more heavily armed than reconnaissance patrols, combat patrols have a mission to capture enemy documents, provide security, and capture or destroy enemy equipment and installations. Such action is ordinarily followed by a return to friendly positions. Regardless of the mission, the patrol reports any information concerning the enemy and terrain acquired during the accomplishment of the assigned mission. There are four types of combat patrols: raid, contact, ambush, and security (normally conducted by a Marine rifle platoon). A rifle platoon reinforced with crew-served weapons is normally considered the minimum size for contact, economy of force or ambush patrols. In some situations, such as the capture of a small enemy outpost, a rifle platoon could conduct a raid. However, a raid is a complex mission and, due to the organization of a raid force (command, reconnaissance, assault, support, security, and reserve elements), a rifle company is normally the smallest force assigned to a raid.

Classification as to Means of Movement

Foot Patrols

Movement by foot is the most common means; however, there are inherent disadvantages. Foot patrols travel slowly and carry limited quantity and types of equipment and supplies. Range and area coverage is relatively restricted. Foot patrols also have apparent advantages in that they have fewer restrictions as to terrain that can be covered; are more difficult for the enemy to detect; provide thorough coverage within limits of range; and are generally not inhibited by weather.

Motorized Patrols

Where terrain and road networks permit, a motorized patrol overcomes the inherent disadvantages of the foot patrol. Mechanized forces require patrolling units that can keep pace with them. However, motorized patrols are restricted to certain types of terrain, and tend to bypass areas that may be advantageous to and occupied by enemy infantry.

Waterborne Patrols

Waterborne patrols move over seas, lakes, rivers and streams, canals, and other inland waterways. The water is either used as a medium of entry to an objective area or is the actual patrol route. Waterborne patrols are limited by the location of water routes in the terrain and tend to bypass areas that may be advantageous to and occupied by the enemy.

Helicopterborne Patrols

Where terrain is extremely difficult or the enemy situation precludes the use of vehicle or motorized patrols, helicopterborne patrols are a method or means to conduct a patrol.

8005. TRAINING

Training is essential to successful patrolling. Premature and unordered actions by members of the patrol destroy coordination and control. Leaders are trained to issue their orders calmly to inspire confidence and discipline, and to avoid misunderstanding. Patrol members must work together and fight as a team. Training should develop the following skills:

- Expertise in handling individual and special weapons, and familiarity with enemy weapons that may be captured.
- Recognize camouflaged personnel, equipment, and defensive positions; ability to pick up fleeting targets fire the rifle from any firing position.
- Understand fire discipline and, after weapon firing, immediately change position (see fig. 8-1).
- Quick and accurate observation skills, and the ability to recall and transmit clearly and briefly, both orally and in writing.

Figure 8-1. Changing Position After Firing.

- Recognize and quickly respond to improvised signals, visually or by sound.
- Ability to swim with weapon and equipment.
- Use issued or improvised camouflage suits and garnish helmet in order to blend with the surroundings. Smudge face, hands, and any bright surfaces of weapons and equipment with some substance, such as mud or charcoal, to prevent the reflection of light.
- Silence self, equipment, and weapon.
- Use antimalarial and water purification tablets.
- Acclimation to temperature extremes.
- Develop a sense of direction and learn how to follow a course by compass, stars, sun, flow of streams, prominent terrain features, and by observing other natural phenomenon. Learn to determine the distance traveled from a known point and to keep a record of azimuths and the distance traveled on each azimuth (dead reckoning).
- Call for and adjust indirect fire assets.
- Familiarization with all communications assets and the use of field expedient antennae.

8006. KEYS TO SUCCESSFUL PATROLLING

Regardless of the category or means of conducting a patrol, the keys to successful patrolling are—

- Detailed planning. Every portion of the patrol must be planned, all possible contingencies considered.
- Productive, realistic rehearsals. Each phase of the patrol is rehearsed, beginning with actions in the objective area. Similar terrain and environmental conditions are used when conducting rehearsals.
- Thorough reconnaissance. Ideally, the patrol leader will physically conduct a reconnaissance of the route and objective. Photographs and/or maps will be used to supplement the reconnaissance.
- Positive control. The patrol leader must maintain positive control, this includes supervision during patrol preparations.
- All-around security. Security must be maintained at all times, particularly near the end of the patrol where there is a natural tendency to relax.

Chapter 9. Patrol Organization

Organizing a patrol is a two-step process: the general organization of the entire patrol and the task organization of various patrol elements. Normally, the nature of patrolling does not permit long preparation periods and rehearsals to specifically build a unit for each mission. Accordingly, the patrol leader must combine unit integrity considerations with proven concepts of patrol organization.

9001. General Organization

The patrol leader establishes a patrol headquarters and elements to accomplish the mission.

The headquarters is composed of the patrol leader and the personnel who provide support for the entire patrol, such as a forward observer, corpsman, and radio operator.

The major subdivisions of reconnaissance and combat patrols are elements. The existing infantry structure (squads and fire teams) is reinforced as required. For example, a reinforced platoon tasked to conduct a combat patrol that will raid an enemy outpost could be organized as follows:

- Platoon headquarters (command element):
  - Patrol leader (platoon leader).
  - Assistant patrol leader (platoon sergeant).
  - Navigator.
  - Radio operator (company tactical net).
  - Radio operator (patrol tactical net).
  - Corpsman.
- The first squad (security element) provides security en route to the objective area (the point, flank security, and rear security) and at the objective area (the flanks and the objective rally point).
- The second squad (support element) supports by fire for the attack, covering fire for the withdrawal, and supporting fires to cover the crossing of danger areas.
- The third squad (assault element) provides the assault force to attack and seize the objective; searchers to clear the objective; pacers, compass man, navigator, and the assistant patrol leader en route and back from the objective area.

Any attachments a patrol may have (i.e., demolition team, scout snipers, and machine gun squad) will be added to the element that supports its function. For example, the demolition personnel should go with the unit conducting the attack, and scout snipers and machine gunners should stay with the support squad.

9002. Task Organization

The preceding paragraphs described the elements necessary for a patrol to accomplish its mission. These elements reflect the internal functions or tasks required for the patrol to succeed. Depending on the METT-T, there are various methods of grouping these elements together. Task-organization is the further subdivision of patrol elements into teams that are required to perform essential tasks. In creating teams, unit integrity of infantry units should be maintained.

The patrol is organized so each individual, team, and element is assigned a specific task, but capable and prepared to perform other tasks. This may not be possible for certain specialist tasks requiring a trained technician.

Chapter 10. Patrol Preparation

For a patrol to succeed, all members must be well trained, briefed, and rehearsed. The patrol leader must have a complete understanding of the mission and a thorough understanding of the enemy and friendly situations. The patrol leader should make a complete reconnaissance of the terrain to be covered (either visual or map), and must issue an order to the patrol, supervise preparations, and conduct rehearsals.

10001. MISSION

The mission assigned to a patrol must be clear and oriented toward one objective with a specific task and purpose. More than one primary objective or indefinite missions invites confusion, casualties, and failure.

10002. FACTORS INFLUENCING PATROL SIZE

The circumstances under which patrolling by infantry units is conducted make it necessary for combat patrols to be able to fight offensively, security patrols to defend themselves, and reconnaissance patrols to move quickly and only fight if necessary to break contact or defend themselves. The size of a patrol depends on METT-T.

Generally, a patrol should contain the least number of members needed to accomplish the mission. Combat missions ordinarily require larger patrols than reconnaissance missions.

10003. COMMANDER'S DUTIES

Determining Patrol Requirements

The need for conducting patrols derives from the commander's stated mission (issued by higher level commander) and other specified and implied tasks (secondary or supporting efforts necessary to accomplish the stated mission). This requires analyzing the unit's mission and determining the necessary reconnaissance and/or combat tasks that must be performed. By considering the mission, along with time available, the commander develops the overall concept of operations to include the patrol plan and the specific tasks for each patrol.

Assigning Units

When assigning patrol missions, maintaining the integrity of the existing unit while considering the skills and experience of the unit and its leader are critical factors to the infantry commander. To provide operational depth and equitable apportionment of hazardous assignments, the commander ensures that each of the subordinate leaders and units develops the skills and experience necessary to conduct successful patrols. Assignment of patrol units must consider the commander's concept of operations as a whole and the plans for subsequent employment of assigned forces after completion of the assigned patrolling mission.

Providing Adequate Time

The commander must allow the patrol sufficient preparation time by completing the mission analysis, estimate, and preparation of orders. The commander should use warning orders to alert subordinates to possible requirements and afford them the opportunity for concurrent planning.

Providing the Patrol Leader Information

The following information should be provided:

- A simple, straightforward explanation of the mission, particularly for night patrols.
- General routes (defined by checkpoints) or exact routes (defined by avenues of approach or other terrain features) to follow.
- Enemy composition, disposition, and strength.
- Location and activities of friendly troops.
- Outposts or other security elements through which the patrol is to pass.
- Terrain conditions.
- Missions and routes of other patrols.

- Time patrol is to depart and return.
- Method of reporting information while on patrol (radio, messenger), place where messengers are to be sent, and place where the patrol leader is to report upon completion of the patrol.
- The challenge and password to be used for exiting and reentering friendly lines.
- Special instructions such as locations to be avoided and essential elements of information higher headquarters is seeking.
- Report barrier/obstacle plan of friendly units if known; support available from friendly unit (e.g., medical evacuation [MEDEVAC], react force).
- Fire support available.

Providing Special Skills and Equipment

There will be situations when the unit assigned to conduct a patrol does not have the necessary technical skills or equipment organic to it to successfully accomplish the patrol mission. In these cases, additions (attachments) are made to the unit, such as—

- Machine gun and/or shoulder-launched, multipurpose assault weapon (SMAW) team and/or squad.
- Forward observers (mortar/artillery).
- Radio operator(s).
- Combat engineers.
- Tracked vehicle crewman to assess the ability to traverse the terrain.
- Corpsman.
- Other personnel (snipers, translators) as required.

Providing Miscellaneous Support

The commander must ensure that the patrol leader is provided with the food, water, ammunition, radios and batteries, maps, special clothing, and any other items required by the unit (includin attachments) needed for the mission. Post-patrol support such as debriefings by intelligence personnel must also be planned.

Reviewing the Patrol Leader's Plan and Preparations

Once the patrol leader receives the mission, conducts visual and/or map reconnaissance, and develops the plan, the commander confirms the patrol leader's understanding of the mission and plan for accomplishing it. This discussion between the patrol leader and commander ensures that the patrol leader understands the commander's intent and is properly prepared to provide the patrol leader an opportunity to ask for clarification or additional support, if required.

Debriefing the Patrol

Upon return of the patrol, the commander receives the patrol report at a debriefing attended by the patrol leader and all patrol members. The debriefing should be conducted as soon as possible following the patrol's return, while information is still fresh in the minds of the patrol members. A patrol report, based on the information collected during the debrief, is generated and forwarded to the next higher com-mander.

10004. PATROL LEADER DUTIES

The patrol leader organizes and prepares the patrol by using the six troop-leading steps to make the best use of resources available. These steps are to—

1. Begin planning.
2. Arrange for reconnaissance and coordination.
3. Make reconnaissance.
4. Complete the plan.
5. Issue the order.
6. Supervise.

Begin Planning

The patrol leader begins by evaluating all factors affecting the mission. He looks for possible courses of action that lead to a decision, and then transforms this decision into an order. The first step includes making an initial assessment and decision on using available time, issuing a warning order and initial preparatory tasks, and initiating his estimate.

The initial planning effort assesses the time, assistance, and information available, and plans the proper use of each. Time allowances include reconnaissance; completion of the estimate and order; troop preparation; and such briefings, rehearsals, and inspections as required before beginning the patrol.

The patrol leader reviews the mission and the attachments and/or support available and decides what preparatory efforts must begin immediately. Proper use of subordinates to manage these initial tasks during this period reduces preparation time and frees the patrol leader for proper planning and reconnaissance. A warning order is issued using a modified five-paragraph order format (situation, mission, execution, administration and logistics, and command and signal [SMEAC]) as a checklist. (Refer to app. A for specific information that a patrol warning order should contain within each checklist item.

Once the initial preparations are set in motion, the patrol leader begins estimating by analyzing the mission; considering the friendly, enemy, and operating environments; considering each course of action available against what the enemy might do; comparing the courses of action in terms of mission accomplishment, capabilities, and probable casualties; and choosing one that becomes the basis for concept and order. While the patrol leader should use and organize notes, estimates must be done quickly and accurately, particularly for immediate situations. To organize thinking, the patrol leader uses METT-T, which consists of:

- Mission—the mission assigned to the patrol and how it relates to the mission of the commander who is sending the patrol.
- Enemy—what is known or suspected of enemy presence and capabilities, habits and characteristics, and fighting techniques.
- Terrain and weather—including ground, vegetation, drainage, weather, and visibility.
- Troops and support available—friendly situation and support available.
- Time available—the constraints and impact of time on preparation and mission accomplishment.

The estimate begins with mission analysis, which is the most important part of the entire planning process. Here, the patrol leader considers the specified tasks of the mission assigned and identifies other significant actions (specified and implied tasks) that must be undertaken to accomplish the stated mission. These tasks are arranged in sequence of accomplishment. The tasks and sequence create the framework for developing courses of action for the patrol concept of operations.

Arrange for Reconnaissance and Coordination

The patrol leader arranges a personal reconnaissance to observe as far forward as possible and also coordinates with the appropriate commanders for the patrol's "passage of lines" (see paragraph 11001 for complete definition) and supporting fires. The patrol leader also coordinates with other patrol leaders who may be operating in the same or adjacent areas and requests that the commander assigning him the patrol mission coordinate the patrol action with adjacent commanders, local security, and night defensive fires, as appropriate. The patrol leader may delegate any or all of these arrangements to the assistant patrol leader if the patrol leader requires the time for planning.

Make Reconnaissance and Complete the Estimate

The patrol leader uses personal reconnaissance to answer questions that arise from the map reconnaissance and METT-T evaluation. Specific points include passage points, lanes through obstacles, locations of friendly listening posts and observation posts, possible approach and return routes, enemy positions (if any), and intermediate observation points on the way to the objective.

In selecting approach and return routes, the patrol leader chooses routes that best use concealment and avoid opposition and obstacles. To lessen the chances of ambush by the enemy, the return trip is planned along a different route. In addition to personal reconnaissance and review of the map and aerial photographs, the advice of other patrol leaders who already are familiar with the terrain and the objective area should be considered.

After compiling information about the situation and possible time constraint, the patrol leader completes an estimate. The first step is developing courses of action, each of which will provide for movement to the objective area, mission accomplishment, and the return, based on the tasks and their sequencing identified in mission analysis. While the eventual concept of operations is presented in order of occurrence, the patrol leader must develop the courses of action by either backward or forward planning. In situations where the objective is well defined and there is sufficient information to plan the action for mission accomplishment (reconnaissance or combat), the patrol leader begins the scheme for accomplishing the mission at the objective and then, planning backwards, considers the options for getting there and back.

The following sections on movement to and return from the objective area, reconnaissance missions, and combat missions discuss methods and options available to the patrol leader in developing the courses of action. The principal variables between courses of action will be who, where, and how in the following:

- Patrol task organization.
- Routes to the objective area.
- Observation point(s) (reconnaissance patrol), ambush site, form of maneuver, type of ambush (combat patrol), fire support plan.
- Return routes.

The patrol leader then mentally considers the progress of each course of action (a map or simple sketch is a useful aid) against expected and unexpected enemy action. By comparing the options against each other and prospective enemy opposition, the patrol leader chooses the course of action that has the best chance of success. Includes in this mental preview process is the time to determine the patrol plan for unexpected contingencies (enemy attack/counterattack, casualty handling). These contingency actions, together with the selected course of action, become the patrol concept of operations.

Once the patrol leader determines the scheme of maneuver, the fire support required to accomplish the mission is addressed including the fire power organic to the patrol and what additional indirect fire support will have to be provided by other units. When planning for indirect fire support, the patrol leader considers the following questions:

- Will artillery, mortar, or close air support be required at the objective area (combat patrols)?
- What artillery and mortar targets exist along the routes to and from the objective area that can be employed by the patrol if it encounters the enemy during movement (reconnaissance patrols and combat patrols)?
- What additional fire support will be required to cover the patrol's movement from the objective area back to the friendly area once the enemy is aware of the patrol's actions at the objective area (combat patrols)?

The effect that casualties have upon the patrol depends upon many factors. Generally, more casualties can be expected in a combat patrol than in a reconnaissance patrol. A patrol may continue on to the objective carrying its casualties, send them back with a detail of Marines, abort the mission and return the entire patrol with the casualties, or call their parent unit for assistance.

Some factors that determine what action the patrol leader takes are: patrol's mission; unit's standing operating procedure for handling wounded; number of casualties and nature of their injuries; availability of aid, helicopters or other means of casualty evacuation. Helicopter evacuation should only be used for the most serious casualties. For infantry units conducting patrols in proximity to the enemy, helicopter evacuation of casualties may compromise the patrol's mission and force the patrol to return to friendly positions before the mission is completed.

The patrol leader determines the requirement for nuclear, biological, and chemical (NBC) defense equipment. Gas masks should always be carried due to the availability of riot control agents (RCAs) to the enemy. If chemical or biological agents have been employed in the area that the patrol must pass through, protective garments will have to be worn by patrol members for part of or the entire patrol. Wearing extra clothing and carrying extra equipment affects the speed of the patrol's movement. A contingency plan for post patrol decontamination must be developed.

Complete the Plan

At this point, the patrol leader has completed the basic thinking necessary for accomplishing the assigned mission. The patrol leader prepares the patrol order to spell out the details, assign tasks to subordinates, and explain the entire endeavor for ease of understanding by the other members of the patrol.

Prepare the Order

The patrol leader's order contains more detailed information than discussed in the warning order. Orders follow the prescribed five-paragraph order format but contain greater detail. The patrol order is a modified 5-paragraph order; the major modifications are to paragraphs 3a and 3c. The format for the order is contained in appendix B.

Issue the Order

The patrol leader asks for a status report on the initial preparatory tasks assigned to subordinate leaders and specialists when the warning order was issued. When the patrol leader has completed planning and initial preparations have progressed to the point where the patrol order may be issued, the members of the patrol are assembled. Roll call is taken to ensure all patrol members are present, then the prepared order is issued.

This will be the only opportunity for the patrol leader to issue detailed instructions. The mission, in particular, must be unmistakably clear so that once the patrol is committed, all subordinate leaders can act with unity of purpose.

Whenever possible, the patrol leader should have a Marine, such as the navigator, build a terrain model using dirt, sand, twigs, etc., explaining the concept of operations for movement to the objective area, actions at the objective area, and the return. Terrain models provide patrol members with a clear and simple layout of the area of operations and key terrain.

Supervise

Inspections and rehearsals, vital to proper preparation, are conducted even though the patrol leader and patrol members are well experienced in patrolling. Inspections determine the patrol's state of physical and mental readiness.

The patrol leader inspects before rehearsals to ensure completeness and correctness of uniform and equipment. The following areas are checked:

- Camouflage.
- Identification tags, Geneva Convention cards.
- Prescribed equipment, weapons, and ammunition are available and serviceable.
- Tape and other items are used to "silence" equipment (prevent noise produced during movement).
- Items that could provide information to the enemy (e.g., letters and papers) remain behind.
- Unnecessary equipment and excess weight remain behind.

The patrol leader questions each patrol member to ensure the following is known:

- The mission, planned routes (primary and alternate), and the fire support plan of the patrol.
- The individual's role: what to do and when to do it.
- What others are to do and how their actions impact.
- Challenges and passwords, codes, reporting times, radio call signs, frequencies, and any other pertinent details.

There is usually a period of time between final rehearsal and departure. The patrol leader reinspects just before departure to ensure all equipment is still in working order and the unit is ready to embark on the mission.

Rehearsals ensure the operational proficiency of the patrol. Plans are checked and needed changes are made. The patrol leader verifies the suitability of equipment. It is through rehearsals that patrol members become thoroughly familiar with the actions to take during the patrol.

If the patrol will operate at night, both day and night rehearsals are conducted. Terrain similar to that over which the patrol will operate is used. All actions are rehearsed. If time is limited, the most critical phases are rehearsed. Action at the objective is the most critical phase of the patrol and is always rehearsed.

An effective method is to talk the patrol through each phase, describing the actions and having each member perform individual duties. When satisfied, the patrol leader walks the patrol through all phases of the patrol using only the signals and commands to be used during the actual conduct of the patrol. Rehearsals continue until the patrol is thoroughly familiar with the plan. The rehearsal is also used to test the soundness of the patrol order and patrol organization.

After the rehearsal, the patrol leader makes final adjustments to the plan and patrol organization based on what was learned during the rehearsal and from other sources, such as the S-2 and adjacent patrols. When this is completed, the patrol leader issues final instructions to subordinate leaders noting any changes made in the patrol organization or plan. While the subordinate leaders are briefing the remainder of the patrol members, the patrol leader reports to the commander stating that the patrol is ready to begin the mission. The patrol leader also coordinates the location and time that the patrol can test fire all weapons prior to departure.

CHAPTER 11. MOVEMENT TO AND RETURN FROM THE OBJECTIVE AREA

This chapter provides guidance to patrol leaders for movement to and return from the objective area. Action in the objective area depends on whether the patrol is assigned a reconnaissance mission, combat mission or security mission. Chapters 12 and 13 provide detailed guidance on these types of missions.

11001. PASSAGE OF LINES

A passage of lines is an operation in which a force moves forward or rearward through another force's combat positions with the intention of moving into or out of contact with the enemy. (JP 1-02)

During the initial preparation for the patrol, the patrol leader selects a patrol assembly area and reconnoiters the area of passage designated by the commander. In coordination with the unit commander responsible for the area of passage, the patrol leader identifies gaps or lanes in minefields and wire obstacles and locates local security elements through which the patrol will pass. The patrol leader also checks the route from the patrol assembly area to the passage point or contact point where the patrol will depart friendly lines. If possible, both the route to the passage point and the route through the frontlines should be concealed from the enemy's view.

The patrol leader also reconnoiters the area for return passage of lines and coordinates with the unit commander responsible for the area of passage for passage points and lanes as necessary. The patrol leader observes these points from the direction that the patrol will use upon return to friendly lines, if possible, to aid in recognition upon return. The patrol leader provides the forward unit with information about the size of the patrol, general route, and expected time of return. The manner of challenge and recognition of the returning patrol should be coordinated in detail.

Upon return to friendly forward local security squads and/or frontlines, the patrol leader leaves the patrol in a covered position and moves forward with a radio operator and at least one Marine for security to make contact with the friendly unit in the manner agreed. After contact is made and recognized, the patrol leader rejoins the patrol and takes them to the passage point, personally checking in each member.

11002. ORGANIZATION FOR MOVEMENT

The patrol's task organization establishes the elements and teams needed to accomplish the mission in the objective area and to and from the objective. The patrol leader determines the formation(s) in which the patrol moves to the objective area.

Formations

The proper use of patrol formations is critical to the patrol's success. The squad and fire team formations described in FMFM 6-5 (proposed MCWP 3-11.2), *Marine Rifle Squad*, also apply to infantry patrolling formations. Because the movement of the patrol must be concealed from the enemy, the patrol normally moves through terrain that provides concealment. Control of the patrol in this type of terrain is difficult; thus the column formation, which is easily controlled, is normally used. However, as various types of terrain are encountered, the patrol leader uses the same considerations in determining the appropriate formation used in other infantry operations.

The standard squad and fire team formations are adaptable to a patrol. The patrol may change formations en route to match the situation and terrain. The patrol leader may have to sacrifice some control for better dispersion or give up some speed for greater stealth and more security. Other considerations include—

- Visibility, weather, terrain, and vegetation will influence dispersion and control of individuals and units. These factors may also affect the enemy; if visibility is good for the patrol, it is also good for the enemy. Two pieces of luminous tape worn on the back of the collar will aid in control and movement on dark nights. The collar is turned down when near the enemy. The tape can also be worn on the back of the cap, but should be covered or removed when near the enemy.

- Preserving the integrity of fire units (fire teams and, if attached, machine-gun teams) is of primary importance. If team members are detached from a machine-gun team, the unit integrity is lost and effectiveness is reduced. The patrol leader must also position fire units so as not to mask their fires.
- All-around defense of the patrol must not be sacrificed under any circumstances. The conventional squad and platoon formations provide adequate firepower in any direction required. When attachments are made to the patrol, the attachments are positioned within the formation to enhance the firepower of the patrol. If a fire unit, such as a machine-gun team or squad, is attached, it is incorporated into the all-around defense of the patrol by modifying the conventional formation (see fig. 11-1). It is permissible to employ machine-gun teams individually during movement. However, assault rocket squads and teams should be employed primarily as rifle fire units for movement. Firing the SMAW or AT-4 from within a patrol formation can present a serious back-blast danger to patrol members and should only be fired on the patrol leader's direction.
- Time allotted for mission accomplishment is also a major consideration. In selecting the formations, the patrol leader must consider the speed of the movement required to meet the time constraints (if any) imposed on the patrol. If required to meet a time schedule, a formation that permits rapid movement should be used. Speed, however, must never be permitted to force the patrol leader to make rash tactical decisions.

Exercise of Control

The patrol leader is positioned for best control over the patrol. The assistant patrol leader moves at or near the rear of the patrol. Other subordinate leaders move with their elements. All patrol members assist by staying alert and passing on signals and orders. A signal to halt may be given by any patrol member, but the signal to resume movement is given only by the patrol leader.

Arm-and-hand signals are the primary means of communication within a patrol and should be used exclusively when near the enemy. All members must know the standard infantry signals (refer to FMFM 6-5), as well as any special signals required, and be alert to receive and pass them to other members.

The patrol leader should speak just loudly enough to be heard. At night, or when close to the enemy, the patrol leader halts the patrol and has subordinate leaders come forward. They speak in a low voice and then pass the information to their subordinates by moving from member to member.

Radios provide a means of positive control within a large patrol; however, radios should be used only when arm-and-hand signals or face-to-face contact between the patrol leader and subordinate leaders is impractical. When close to the enemy, words are

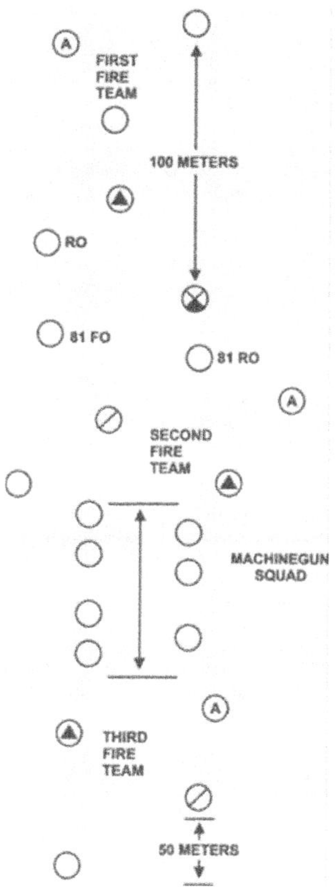

Figure 11-1. Squad Patrol with Attached Mortar Forward Observer Team and Machine Gun Squad.

spoken into the microphone with hands cupped over it in a low voice.

Other sound signals may be used if the patrol leader is sure they serve the purpose intended. Planned sound signals are rehearsed before starting on the patrol. Sound signals used must be simple, natural sounds that are few in number and easily understood. Bird and animal calls are seldom satisfactory.

Infrared equipment, such as the sniper scope and infrared filters for the flashlight, may be used as means of sending and receiving signals and maintaining control at night.

Luminous tape may be used to assist in control at night. Small strips on the back of the cap or collar of patrol members aid in keeping visual contact with the front member. However, the luminous tape must be covered when near the enemy.

An important aspect of control is accounting of personnel, especially after crossing danger areas, enemy contacts, halts, and exiting and re-entering friendly lines. The patrol leader may arrange for the last member to send up the count automatically after crossing danger areas, enemy contact, and halts. In large patrols or when moving in a formation other than a column, subordinate leaders check members and report the count to the patrol leader by the quickest method appropriate to the circumstances.

Navigation

One or more members are assigned as navigators for the patrol. Their function is to assist the patrol leader in maintaining direction by the use of the compass.

The patrol leader assigns at least two members as pacers to keep track of the distance from point to point. The average of their count is used to approximate the distance traveled. Pacers are separated so they do not influence each other's count. Pacers must know how to convert their own paces to meters.

The route is divided into increments, with each increment starting at a recognizable point on the ground. The pacers begin their counts from zero at the beginning of each leg. This makes the pace count easier to keep and checks for accuracy.

The pace count is sent forward when the patrol leader turns to the member behind and in a low voice says, "send up the pace" or uses the arm-and-hand signal of tapping his boot. This signal is passed to both pacers, who in turn send up the pace count in meters; for example, "two-hundred" or "one-seven-five."

Patrol members must understand that the counts of both pacers are sent forward. The patrol leader must know the counts of both men in order to check them.

Security

The patrol is dispersed consistent with control, visibility, cover, and concealment. Scouts are employed to the front, flanks, and rear of the patrol to provide security. For the patrol members traveling in the main body of the patrol, areas of responsibility are assigned to the front, flanks, and rear. Scouts are the eyes and ears of the patrol leader. They move when and where directed by the patrol leader and maintain contact with the patrol leader at all times, except when momentarily obscured by vegetation or other terrain features.

Front

Small patrols (squad size) may employ from one scout up to a fire team as the point, depending on the enemy situation, terrain, and route being followed by the patrol. Normally, squad-size patrols will use two scouts as the point. The point is responsible for investigating the route of advance immediately to the front of the patrol.

The point moves as far ahead of the patrol as visibility and terrain permit. When visibility is good, the point may precede the main body by as much as 100 meters. The point travels right and left ahead of the patrol, searching the area over which the patrol will pass.

The point maintains direction by knowledge of the general route to be followed and visual contact with the patrol leader. The patrol leader or the navigator ensures that the point is proceeding correctly.

The point, which stays far enough ahead of the patrol to provide security, is not a trail breaker for the patrol. If the point loses contact with the patrol, the point waits for the main body to catch up or moves rearward if contact is not quickly regained.

One of the navigators may be positioned with the point. One or more members works as the point while the other is the navigator.

Flank

Flank security for a patrol of squad size or less may be provided by using one or two members on either flank. If two scouts are assigned to a flank, one is positioned to observe the patrol leader and the other works farther out from the patrol. The scout who must observe the patrol leader remains within a maximum distance of 100 meters. The scout farther out remains in sight of the inside scout but normally does not move more than 20 or 25 meters away and remains prepared to relieve flank security regularly. Moving through dense woods or jungle may render the use of flank security impractical because of reduced visibility. In such cases, it moves with the patrol itself, but maintains observation to its assigned flanks.

Rear

A small patrol normally has only one rifleman assigned as rear security. An interval between the member assigned as rear security and the last Marine of the patrol is maintained at the limit of visibility, up to 50 meters. This member maintains rear security for the patrol by constantly observing to the rear.

Halting

Speed of movement is slower at night than in day patrols and reduces the danger of a Marine becoming separated from the patrol. The patrol occasionally halts to observe and listen for enemy activity; this is called a security halt. Upon signal, when reaching a danger area and periodically throughout movement en route, every member freezes in place, remains quiet, observes, and listens. It may be necessary to call a security halt just after departing friendly areas and just before entering friendly areas.

The patrol may halt briefly to send a message, eat, rest, check direction, or make a reconnaissance. The area selected should provide adequate concealment and cover, as well as favor the defense. All-round security is established and the patrol leader ensures all members move out when the patrol resumes movement. (For extended halts, see paragraph 11005.)

Infiltration

The disposition of enemy forces may sometimes prevent a patrol from entering the enemy occupied area as a unit; however, pairs of scouts or fire teams may slip through without being discovered. (Refer to FMFM 6-5 for infiltration techniques and procedures).

11003. CONTROL MEASURES FOR MOVEMENT

Checkpoints

A checkpoint is a predetermined point on the surface of the Earth used as a means of controlling movement, a registration target for fire adjustment or reference for location (JP 1-02). Checkpoints are a means of control between the parent unit and the patrol. These locations are decided upon and coordinated before the patrol leaves, so that both the patrol members and parent unit will know the patrol's location when it reports in. The parent unit can follow the progress of the patrol without transmitting coordinates over the radio that the enemy could monitor.

Rally Points

A rally point is an easily identifiable point on the ground where units can reassemble and reorganize if they become dispersed (MCRP 5-12C). It should provide cover and concealment, be defensible for at least a short time, and be easily recognized and known to all patrol members. All rally points are considered and identified as tentative rally points until they are occupied, found to be suitable, and designated as rally points by the patrol leader.

Initial Rally Point

The initial rally point is within friendly lines where the patrol can rally if it becomes separated before departing the friendly area or before reaching the first en route rally point. It may be the assembly area where the patrol waits while the patrol leader contacts the last friendly position through which the patrol will pass. The location of the initial rally point must be coordinated with the forward unit commander.

En route Rally Points

En route rally points are between the initial rally point and the objective rally point; and from the objective rally point back to the point where the patrol reenters friendly lines. They are determined as the patrol passes through a likely area that is suitable for a rally point.

Objective Rally Point

The objective rally point is located nearest the objective where the patrol makes final preparations prior to approaching the objective. It also serves as a location where the patrol reassembles after completing actions on the objective. The objective rally point must be suitable to accommodate those activities accomplished prior to actions on the objective. This position must provide the patrol concealment from enemy observation and, if possible, cover from enemy fires. It may be located short of, to a flank or beyond the objective. It should be out of sight, sound, and small-arms range of the objective area. The patrol leader's reconnaissance of the objective is made from this position; it is the release point from which patrol elements and teams move to the objective to accomplish the mission.

Rally Point Selection

The patrol leader selects likely locations for tentative rally points during reconnaissance or map study. A tentative rally point must be confirmed and announced after examination proves suitability. A tentative initial rally point and a tentative objective rally point are always selected and identified in the patrol order. If necessary, the patrol leader selects additional rally points en route as suitable locations are reached. When the patrol reaches a danger area that cannot be bypassed, such as an open meadow or stream, the patrol leader selects a rally point on both the near and far side. If good locations are not available, the patrol leader designates the rally points in relation to the danger area. The patrol leader will say, for example, "50 meters this side of the trail" or "50 meters beyond the stream."

Rally Point Use

The initial rally point and en route rally points are designated to enable the patrol to reassemble if it is unavoidably separated or dispersed. Identifying features are pointed out. The patrol leader ensures that the information is passed to all patrol members. When crossing a danger area, a near side rally point and a far side rendezvous point are designated. These rally points should only be used when all other methods of retaining control of the patrol have failed. The success of the patrol may be jeopardized if it is dispersed and cannot rally expeditiously.

If the patrol has left the friendly area and becomes dispersed, patrol members return to the last designated rally point (the initial or an en route rally point) unless the patrol leader gives other instructions.

As previously noted, the patrol leader selects two rally points at the near and far sides of danger areas that cannot be bypassed. If the patrol becomes separated or dispersed at a danger area, and there has been no enemy contact, the patrol should reassemble at the rally point on the far side of it. If, however, the patrol is separated or dispersed at a danger area as a result of enemy contact, members who have already crossed the danger area assemble and reorganize at the rally point on the far side; members who have not crossed assemble and reorganize at the rally point on the near side. In the absence of the patrol leader and unless directed otherwise, the senior member at the rally point on the near side takes charge, attempts to move the rallied patrol members to the rally point on the far side, and rejoins the remainder of the patrol.

Rally Point Actions

The patrol leader plans the actions to be taken at rally points and instructs the patrol accordingly in the patrol order. Planned actions at the initial rally point and en route rally points must provide for the continuation of the patrol as long as there is a reasonable chance to accomplish the mission. Plans for actions at rally points should provide for—

- Recognition signals for assembly at rally points.
- Minimum number of members and maximum amount of waiting time required before the senior member at the rally point moves the rallied patrol members onward toward the objective or returns to friendly lines.
- Instructions for patrol members who find themselves alone at a rally point.

11004. PRECAUTIONS AT DANGER AREAS

A danger area is any place where the patrol is vulnerable to enemy observation or fire (open areas, roads, trails, and obstacles such as barbed wire, minefields, rivers and streams, and lakes). Any known or suspected enemy position the patrol must pass is also a danger area. The patrol leader plans for crossing each danger area and includes these plans in the order.

The patrol reconnoiters the near side of a danger area first, then the patrol leader sends scouts to reconnoiter the far side. Once the scouts report that the far side is clear of the enemy, the remainder of the patrol crosses the danger area. As each individual or group crosses the danger area, they are covered by those remaining and by those who have successfully crossed. Enemy obstacles are avoided since they are usually covered by fire.

In crossing a river, the near bank is reconnoitered first; then the patrol is positioned to cover the far bank. Scouts are sent across to the far bank. After the far bank has been reconnoitered and the scouts report that it is clear of the enemy, the patrol crosses as rapidly as possible. This may be done individually or in pairs. If crossing the river requires swimming, the patrol uses improvised rafts to float equipment, weapons, and ammunition across. (Refer to MCRP 3-02C, *Water Survival Handbook.)*

A road or trail is crossed at or near a bend or where the road is narrow. Observation is restricted and, if the enemy is present, the patrol is exposed as short a time as possible. The near side is reconnoitered first, then scouts are sent across to reconnoiter the far side. This includes reconnoitering the tentative rally point on the far side. Once the scouts report "all clear," the remainder of the patrol crosses rapidly and quietly.

If the patrol must pass close to an enemy position, it takes advantage of battlefield noises to cover the sounds of movement. If supporting fires are available, the patrol leader can call for them to divert the enemy's attention as the patrol passes.

11005. HIDE

When a patrol is required to halt for an extended period in an area not protected by friendly troops, the patrol moves into a location which, by the nature of the surrounding terrain, provides passive security from enemy detection. Such an assembly area is termed a *hide*. To establishment a hide—

- Cease all movement during daylight hours to avoid detection.
- Hide the patrol for an extended period while the patrol leader conducts a detailed reconnaissance of the objective area.
- Rest and reorganize after extended movement.
- Reorganize after a patrol has infiltrated the enemy area in small groups.

The patrol leader's plan must include tentative hide locations when the patrol's mission dictates an extended halt within enemy areas. These tentative locations must be confirmed by actual ground reconnaissance prior to occupation by the patrol. The plan for a hide includes both passive and active security measures.

Passive security measures are—

- Avoid built-up areas.
- Select an area remote from all human habitation.
- Avoid known or suspected enemy positions.
- Avoid ridge lines, topographic crests, valleys, lakes, and streams.
- Avoid roads and trails.
- Avoid open woods and clearings.
- Select areas offering dense vegetation, preferably bushes and trees that spread out close to the ground.

Active security measures—

- Establish security covering all likely avenues of approach into the site.
- Establish communications (wire, radio, signal, runner) with posted security to provide early warning of enemy approach.
- Select an alternate area for occupation if the original hide is compromised or found unsuitable.
- Plan for withdrawal in the event of discovery.
- Establish an alert plan with a certain percent of the personnel awake at all times.
- Organize the elements of the patrol so necessary activities can take place with a minimum amount of movement.

The size of the area physically occupied by a patrol in a hide and the number of security posts required are governed by the terrain, quantity and quality of cover and concealment, and size of the patrol.

If the situation permits, a hide can also be used as the final preparation position and/or objective rally point.

11006. IMMEDIATE ACTIONS UPON ENEMY CONTACT

A patrol may make contact with the enemy at any time. Contact may be through observation, a meeting engagement or ambush. Contact may be visual, in which the patrol sights the enemy but is not itself detected. When this is the case, the patrol leader can decide whether to make or avoid physical contact, basing his decision on the patrol's assigned mission and capability to successfully engage the enemy unit.

When a patrol's assigned mission prohibits physical contact (except that necessary to accomplish the mission), its actions are defensive in nature. Physical contact, if unavoidable, is broken as quickly as possible and the patrol, if still capable, continues its mission.

When a patrol's assigned mission permits or requires it to seek or exploit opportunities for contact (as in the case of a combat patrol), its actions are offensive in nature, immediate, and positive.

In patrolling, contacts (visual or physical) are often unexpected at very close ranges, and short in duration. Effective enemy fire often provides leaders little or no time to fully evaluate situations and issue orders. In these situations, immediate action provides a means for swiftly initiating positive offensive or defensive action, as appropriate.

Two types of physical contact with the enemy are meeting engagement and ambush. Meeting engagement is a combat action that occurs when a moving force, incompletely deployed for battle, engages an enemy at an unexpected time and place. It is an accidental meeting where neither the enemy nor the patrol expect contact and are not specifically prepared to deal with it. An ambush is a surprise attack from a concealed position.

Immediate Actions

Immediate actions are designed to provide swift and positive small unit reaction to visual or physical contact with the enemy. They are simple courses of action in which all Marines are well trained. Minimal signals or commands are required and they are developed as needed for the combat situation. The signals can, in many cases, be initiated by any member of the unit. It is not feasible to attempt to design an immediate action drill to cover every possible situation. It is better to know the immediate action drill for each of a limited number of situations that may occur during a patrol. Arm-and-hand signals associated with immediate actions—such as FREEZE, ENEMY IN SIGHT, and HASTY AMBUSH RIGHT or LEFT—are contained in FMFM 6-5 (proposed MCWP 3-11.2).

Immediate Halt

When the patrol detects the enemy but is not itself detected, the situation requires the immediate, in-place halt of the patrol. The first member visually detecting the enemy gives the silent signal for FREEZE. Every member halts in place, weapon at the ready, and remains absolutely motionless and quiet until further signals or orders are given.

Air Observation and/or Attack

These actions are designed to reduce the danger of detection from aircraft and casualties from air attack.

When an enemy or unidentified aircraft that may detect the patrol is heard or observed, the appropriate immediate action drill is FREEZE. The first member hearing or sighting an aircraft that may be a threat signals FREEZE. Every member halts in place until the patrol leader identifies the aircraft and gives further signals or orders. Members of the patrol must not look up at the aircraft as sunlight can reflect off their faces even when camouflaged.

When an aircraft detects a patrol and makes a low level attack, the immediate action drill air attack is used. The first member sighting an attacking aircraft shouts, "AIRCRAFT," followed by the direction of the incoming attack: FRONT, LEFT, REAR or RIGHT. The patrol moves quickly into line formation, well spread out, at right angles to the aircraft's direction of travel. As each member comes on line, the member hits the ground, using available cover, then positions the body perpendicular to the aircraft's direction of travel, to present the shallowest target possible (see fig. 11-2 on page 11-8). Between attacks (if the aircraft returns or if more than one aircraft attacks), patrol members seek better cover. Attacking aircraft are fired upon only on command of the patrol leader.

Meeting Engagement

Hasty Ambush. This immediate action is used to avoid contact and to prepare to initiate an unplanned ambush on the enemy. It may often be a subsequent action after the command freeze. When the signal HASTY AMBUSH is given (by the point member, patrol leader or another authorized patrol member), the entire patrol moves quickly to the right or left of the line of movement, as indicated by the signal, and takes up the best available concealed firing positions (see fig. 11-3 on page 11-9). The patrol leader initiates the ambush by opening fire and shouting, "FIRE"; thus ensuring initiation of the ambush if the weapon misfires. If the patrol is detected before this, the first member aware of detection initiates the ambush by firing and shouting. The patrol leader may decide not to initiate the ambush in order to avoid contact unless the patrol is detected. When used as an offensive

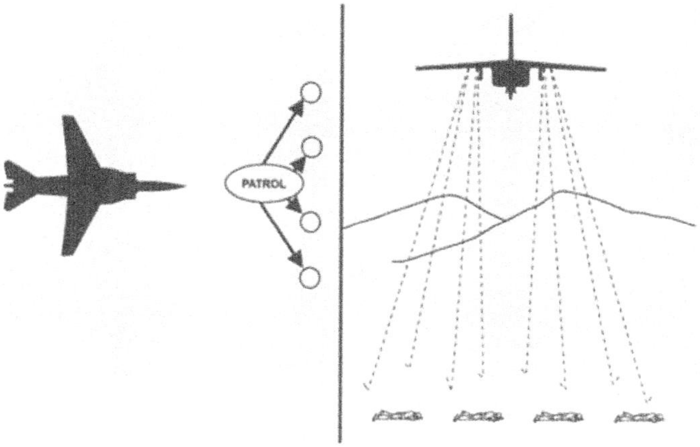

Figure 11-2. Immediate Action, Air Attack.

measure, the enemy is allowed to advance until he is in the most vulnerable position before the ambush is initiated. An alternate means for initiating the ambush is to designate an individual (for example, point or last member) to open fire when a certain portion of the enemy unit reaches or passes that member.

Immediate Assault. This immediate action drill is used defensively to make and quickly break undesired but unavoidable contact (including ambush) and offensively to decisively engage the enemy (including ambush). When used in a meeting engagement, members nearest the enemy open fire and shout, "CONTACT," followed by the direction of the incoming attack: FRONT, LEFT, REAR or RIGHT. The patrol moves swiftly into line formation and assaults (see fig. 11-4).

Defensive Measures

When used defensively, the assault is stopped if the enemy withdraws and contact is broken quickly. If the enemy stands fast, the assault is carried through the enemy positions and movement is continued until contact is broken.

Offensive Measures

When used offensively, the enemy is decisively engaged. Escapees are pursued and destroyed until orders to break contact are given by the patrol leader. If the patrol is fired upon from beyond 50 meters, the patrol must break contact as quickly as possible and continue the mission. If it engages the enemy any longer than necessary to break contact, it may put the mission in jeopardy.

Fire and Maneuver

Fire and maneuver is one means to break contact. One portion of the patrol returns the enemy fire while another portion moves by bounds away from the enemy. Each portion of the patrol covers the other by fire until contact is broken by all.

Clock System

The clock system is another means to break contact. Twelve o'clock is the direction of movement of the patrol. The patrol leader shouts a direction and a distance. For example: "TEN O'CLOCK-TWO HUNDRED," means the patrol should move in the direction of ten o'clock for 200 meters. Patrol members keep their same relative positions as they move so the original formation is not disrupted. Subordinate leaders must be alert to ensure that the members of their elements and teams receive the correct order and move as directed.

Counter Ambush

When a patrol is ambushed, the immediate action drill used depends on whether the ambush is a near ambush

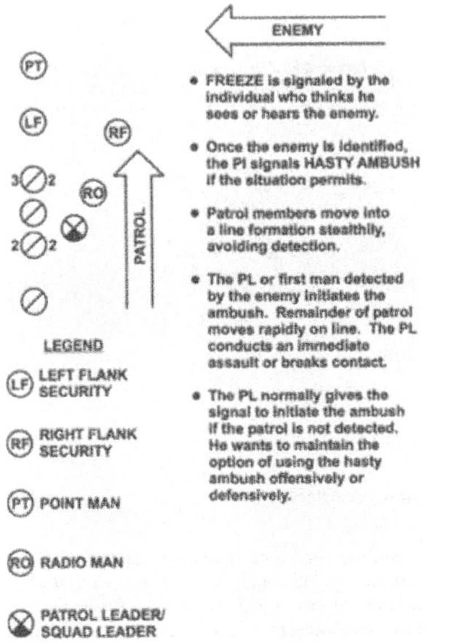

Figure 11-3. Immediate Action, Hasty Ambush.

(the enemy is within 50 meters of the patrol) or a far ambush (the enemy is beyond 50 meters of the patrol). Fifty meters is considered the limit from which the ambushed patrol can effectively launch an assault against the enemy.

Near Ambush. In a near ambush, the killing zone is under very heavy, highly concentrated, close range fires. There is little time or space for members to maneu-ver or seek cover. The longer they remain in the killing zone, the greater the chance they will become casualties. Therefore, if members in the killing zone are attacked by a near ambush, they immediately assault without order or signal directly into the ambush position, occupy it, and continue the assault or break contact, as directed. This action moves them out of the killing zone, prevents other elements of the ambush from firing on them without firing on their own members, and provides positions from which other actions may be taken (see fig. 11-5 on page 11-10). Members not in the killing zone

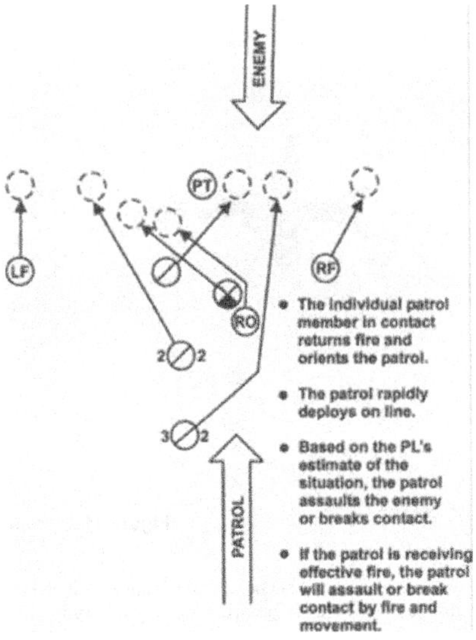

Figure 11-4. Immediate Action, Immediate Assault.

maneuver against the enemy as directed. The assault continues until all patrol members are outside of the killing zone.

Far Ambush. In a far ambush, the killing zone is also under very heavy, highly concentrated fires, but from a greater range. This greater range provides members in the killing zone maneuver space and some opportunity to seek cover at a lesser risk of becoming a casualty. If attacked by a far ambush, members in the killing zone, without order or signal, immediately return fire, take the best available positions, and continue firing until directed otherwise. Members not in the killing zone maneuver against the ambush force, as directed (see fig. 11-6 on page 11-10). The assault is continued against the enemy or until the order to break contact is given.

In each situation, the success of the counter ambush employed depends on the members being well trained in quickly recognizing the distance from which an ambush is initiated and well rehearsed in the proper reaction.

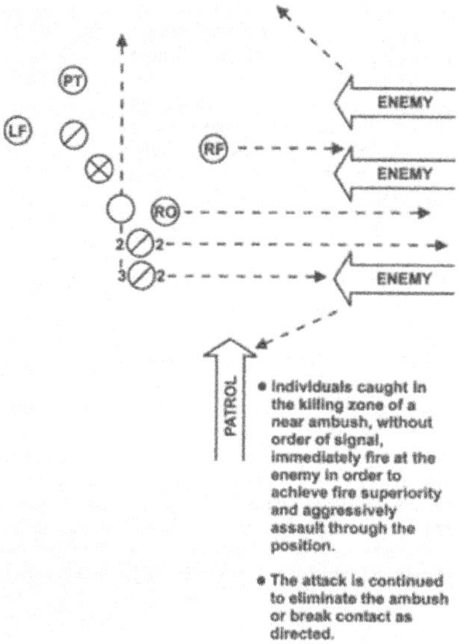

Figure 11-5. Near Ambush.

11007. PATROL LEADER'S ACTION IN A DEVELOPING SITUATION

While good patrolling depends on good planning, the patrol leader's plan must be flexible. Every combat situation develops differently than expected, and the patrol leader must be ready to quickly adapt to the situation as it develops. One of the most difficult tasks in battle is to recognize the correct moment for making a decision. Generally, it is more difficult to determine the moment for making a decision than it is to formulate the decision itself. When the situation demands, decisions must be made promptly without waiting for more information. In a developing situation, the patrol leader should use the following questions as a guide for battlefield decisionmaking:

- How has the situation changed?
- How does the change affect mission accomplishment and the immediate superior's mission?
- Must a decision be made now?
- What are the options?
- Which option best serves the mission and the unit as a whole?
- Which option offers the greatest chance of success?

11008. RETURN FROM OBJECTIVE AREA

After performing actions in the objective area, the patrol reassembles at the objective rally point. This phase of the patrol is perhaps the most difficult and dangerous. Patrol members are experiencing fatigue, emotional letdown, and wounds; they may be low on water and ammunition. Above all, the enemy was likely alerted if not in pursuit. At this point, the patrol leader must move the patrol rapidly but carefully and maintain patrol security at a high level. In returning to friendly lines, the patrol neither uses nor travels near the same route used to get to the objective area. The enemy may have that route covered. For information on reentry of friendly lines, refer to paragraph 11001.

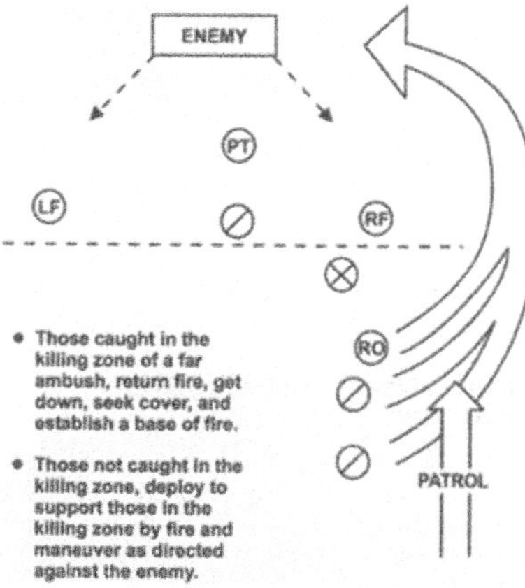

Figure 11-6. Far Ambush.

CHAPTER 12. RECONNAISSANCE PATROLS

The commander needs accurate, timely information about the enemy and the terrain the enemy controls to assist in making tactical decisions. Reconnaissance patrols are one of the most reliable means for obtaining this information. These patrols engage the enemy only when necessary to accomplish their mission or for protection. They generally avoid combat, accomplish the mission by stealth, and do not maintain contact with the unit that sent them out. A reconnaissance patrol is capable of carrying the search for information into the area occupied by enemy forces—usually beyond the range of vision of friendly ground observation posts—and examining objects and events at close range.

12001. GENERAL MISSIONS

General missions for reconnaissance patrols include gathering information about location and characteristics of friendly or hostile positions and installations, terrain (routes, stream crossings), and obstacles.

12002. SPECIFIC MISSIONS

Locate the Enemy

Reconnaissance patrols try to determine the—

- Location of enemy forces, installations, and equipment.
- Identification of enemy units and equipment.
- Strength of enemy forces.
- Disposition of enemy forces.
- Movement of enemy personnel and equipment.
- New or special types of weapons.
- Presence of mechanized units.
- Unusual enemy activity.
- Presence of NBC equipment.

Reconnaissance of Enemy Wire Obstacles

A patrol with the mission of investigating enemy-emplaced wire obstacles employs a formation providing for all-around security and takes precautions against being observed by the enemy. The patrol leader and one Marine inspect each gap and establish its location by means of compass bearings to prominent objects in the rear of enemy or friendly positions or through the use of global positioning satellite receivers.

Reconnaissance of Contaminated Areas

A patrol with the mission of investigating a contaminated area reconnoiters and marks the area's boundaries. Patrol members wear protective clothing and field protective masks. The patrol leader reports the extent of the area, the ttype of agent used, the terrain and vegetation and the method of marking the contaminated area. A sketch of the contaminated area, prepared by the patrol leader, should be included in the patrol report.

Reconnaissance of Enemy Minefields

Patrols assigned to reconnoiter enemy minefields are often composed of specially trained personnel. The type of mine is identified and the possibility of by-passing the area determined.

Reconnaissance of Terrain

The reconnaissance patrol is particularly suited for gathering information about the terrain within their area of operations. (Refer to MCRP 2-15.3B, *Reconnaissance Reports Guide.*) Commanders must know the location and condition of bridges, streams, and roads in order to make proper plans for the movement of troops. The following information should be obtained by a reconnaissance patrol assigned such missions:

- Bridges:
  - Maximum load capacity.
  - Material used in construction (wood, stone, concrete or steel).

  - Material used for abutments (dirt, wood, masonry, concrete).
  - Type and number of supporting members of the bridge.
  - Condition and dimensions of bridge flooring.
  - Presence or absence of mines and demolition charges on the bridge and at entrances/exits.
  - Terrain crossed (ravine, stream, road).
  - Primary use (rail, auto, foot).
  - Location and type of an alternate crossing if bridge is unusable.
- Streams and fords:
  - Width and depth of stream.
  - Composition of the bottom (mud, sand, gravel, rock).
  - Speed of current in miles per hour.
  - Composition and gradient of banks.
  - Surrounding terrain and vegetation cover.
- Roads:
  - Width.
  - Composition of surface.
  - Condition.
  - Road blocks.
  - Gradient/degrees of slopes.
  - Curves (location and length).
  - Pitch and culvert locations.
  - Primary use and frequency of use.

12003. TYPES OF RECONNAISSANCE

Route

Route reconnaissance is a directed effort to obtain detailed information of a specified route and all terrain from which the enemy could influence movement along that route. (Refer to MCRP 5-12C.) It is also described as a form of reconnaissance focused along a specific line of communications—such as a road, railway, or waterway—to provide new or updated information on route conditions and activities along the route. (Refer to MCRP 5-12A, *Operational Terms and Graphics*.)

A route reconnaissance normally precedes the movement of forces. It provides detailed information about a specific route and the surrounding terrain that could be used to influence movement along that route. Considerations include trafficablility, danger areas, critical points, vehicle weight and size limitations, and locations for friendly obstacle emplacement.

The objective area is normally defined by a line of departure, a route, and a limit of advance.

Area

An area reconnaissance is a directed effort to obtain detailed information concerning the terrain or enemy activity within a prescribed area such as a town, ridge line woods or other features critical to operations. (Refer to MCRP 5-12C.) An area reconnaissance could also be made of a single point, such as a bridge or installation. (Refer to MCRP 5-12A.)

Emphasis is placed on reaching the area without being detected. Enemy situations encountered en route are developed only enough to allow the reconnoitering unit to report and bypass.

Zone

A zone reconnaissance is a directed effort to obtain detailed information concerning all routes, obstacles (to include chemical or radiological contamination), terrain, and enemy forces within a zone defined by boundaries. A zone reconnaissance is normally assigned when the enemy situation is vague or when information concerning cross-country trafficability is desired. (Refer to MCRP 5-12C.)

The commander specifies specific routes or areas of interest within the zone. The zone to be reconnoitered usually is defined by a line of departure, lateral boundaries, and a limit of advance.

12004. TASK ORGANIZATION

A reconnaissance patrol is organized around the current structure of the Marine rifle squad with one or more of the squad's fire teams assigned as the reconnaissance element to reconnoiter or maintain surveillance over the objective. At least one fire team acts as a security element whose functions are to—

- Secure the objective rally point (see para. 11003).
- Give early warning of enemy approach.
- Protect the reconnaissance unit.

Reinforcing the squad is based on METT-T. If two squads are required, then the mission should be assigned to a platoon.

A company may send a reconnaissance patrol to specifically reconnoiter a given area (a riverbed or a bridge), or to maintain surveillance on a particular streambed for the next four nights. Reconnaissance patrols perform three basic functions en route to and from the objective:

- Provide control.
- Provide security while moving and in the objective area.
- Conduct reconnaissance or surveillance.

Depending on the size of the reconnaissance patrol, an element must be tasked to reconnoiter the area or zone. In the case of a surveillance mission, two teams are recommended to share the same task (so one could lay up and rest, while the other lays low and hides to maintain strict silence and no movement while surveillance is being maintained on the objective). Whatever the case, each unit that has a surveillance mission or reconnaissance mission must be thoroughly briefed as to what essential elements of information are to be collected as well as their location in the objective area.

The patrol should be organized with one or two fire teams to actually conduct the reconnaissance mission and the remaining fire team to provide security. Each fire team should be prepared to assume either mission. A small area reconnaissance patrol needs only one fire team for the assigned mission. A patrol with a wider area reconnaissance mission should use two fire teams to physically conduct the mission and one fire team for cover and/or security.

The security element for a reconnaissance patrol should be organized to cover the likely avenue of approach into the objective area, to protect the units conducting the reconnaissance, and to cover the objective rally point.

The variations of the special organization for reconnaissance patrols illustrate the patrol leader's flexibility in organizing the patrol to meet mission requirements. The patrol leader is not limited to the variations reflected herein but can choose any special organization as long as that task organization accommodates the requirement for command and control, reconnaissance, and security.

12005. SIZE OF RECONNAISSANCE PATROLS

A reconnaissance patrol should be kept to the minimum number of personnel required to accomplish the mission. A mission requiring a patrol to remain away from its unit for a considerable period of time, or one requiring a patrol to send back information by messenger, increases the size of the patrol. Reconnaissance patrols seldom exceed a squad in strength. Unit integrity should be preserved whenever possible. Intelligence personnel, interpreters, and other specialists, such as radio operators or engineers, are assigned to a patrol if the particular mission requires.

12006. RECONNAISSANCE EQUIPMENT

Patrol members are armed and equipped as necessary for accomplishing the mission. The automatic rifle in each fire team provides a degree of sustained firepower in case of enemy contact. The patrol should have at least two: pairs of binoculars, pairs of wire cutters, maps, compasses, and watches. Night observation equipment may be used. Pencils and small notebooks are carried so notes and sketches can be made. A message book with message blanks and overlay paper is mandatory.

12007. RECONNAISSANCE PATROL ACTIONS AT THE OBJECTIVE AREA

Route

The patrol leader halts and conceals the patrol near the objective area in the final preparation position. The patrol leader then conducts leader's reconnaissance to confirm the plan for positioning the security teams and employing units assigned to the reconnaissance mission. The patrol leader returns to the patrol and positions the security to provide early warning of enemy approach and secure the objective rally point. The reconnaissance unit(s) then reconnoiters the objective area (route). The reconnaissance unit may move to several positions, along or adjacent to the specific route, in order to conduct a thorough reconnaissance. After completing the reconnaissance, each reconnaissance team moves to the objective rally point and reports to the patrol leader. The patrol then returns to friendly lines and the leader makes a full report.

Area

The patrol leader halts and conceals the patrol near the objective area in the final preparation position. The patrol leader then conducts leader's reconnaissance to pinpoint the objective and confirm the plan for positioning the security teams and employing units assigned the reconnaissance mission. The patrol leader returns to the patrol and positions security to provide early warning of enemy approach and secure the objective rally point. The reconnaissance unit(s) then reconnoiters the objective area. The reconnaissance unit may move to several positions, perhaps making a circle around the objective area, in order to conduct a thorough reconnaissance. When the reconnaissance is completed, the patrol leader assembles the patrol and tells members what has been observed and heard. Other patrol members contribute observations. The patrol then returns to friendly lines and the patrol leader makes a full report.

Zone

The patrol leader halts the patrol at the final preparation position, confirms the plan, and conducts leader's reconnaissance. The patrol leader positions the security team and sends out the reconnaissance team. When the entire patrol is used to reconnoiter the zone, it provides its own security. After completing the reconnaissance, each reconnaissance team moves to the objective rally point and reports to the patrol leader. The patrol then returns to friendly lines.

CHAPTER 13. COMBAT PATROLS

Combat patrols are assigned missions that usually include engaging the enemy. They are fighting patrols. Every combat patrol has a secondary mission: gaining information about the enemy and terrain. Combat patrols are employed in both offensive and defensive combat operations and they assist the parent unit in accomplishing its mission by inflicting damage on the enemy; establishing and/or maintaining contact with friendly and enemy forces; denying the enemy access to key terrain; and probing enemy positions to determine the nature and extent of enemy presence.

13001. TASK ORGANIZATION

A combat patrol is organized around the current structure of the Marine rifle platoon. A combat patrol leader should use the unit's normal organization (fire team, squad, and platoon) in assigning functions, patrol missions, and chain of command. Combat patrols must be able to perform the following four basic functions en route to and from the objective:

- Provide control.
- Provide security.
- Provide support by fire.
- Attack or assault the objective.

A rifle platoon could task-organize as a combat patrol as follows:

- Platoon headquarters (patrol headquarters).
- First squad (security).
- Second squad (support).
- Third squad (assault).

Every combat patrol must—

- Provide a control mechanism in the form of a headquarters.
- Designate a unit (a fire team or squad) to provide security while moving en route to the objective or while at the objective. At the objective area, this unit isolates the objective area, secures the objective rally point, and covers the withdrawal.
- Designate a unit to act as support. This unit provides the base of fire in the attack or covers withdrawals or advances.
- Designate a unit(s) to conduct the attack or assault. This unit(s) engages the enemy at the objective area by fire and maneuver or movement. It also operates immediately in the objective area (searching, demolition, prisoners of war, etc.). Paragraph 9001 outlines the general organization of combat patrols.

As in the case with reconnaissance patrols, the task organization of a combat patrol depends on the specific mission assigned. If any special requirements are generated because of the specific mission, the patrol is task-organized to fit the needs of the mission.

13002. EQUIPMENT

Combat patrols are armed and equipped as necessary for accomplishing the mission. In addition to binoculars, wire cutters, compasses, and other equipment generally common to all patrols, it usually carries a high proportion of automatic weapons and grenades. Communications with higher headquarters is important as success of the mission may depend on being able to call for supporting fires. Also, internal radio communications with the units and teams may be useful. However, the patrol must not be so overburdened with equipment as to impede movement or mission accomplishment.

13003. RAID PATROLS

A raid is a surprise attack on an enemy force or installation with the attacking force withdrawing after accomplishing its mission. Raids destroy or capture enemy personnel or equipment, destroy installations, or free friendly personnel who have been captured by the enemy. Patrolling techniques are used in planning and when moveing to and from the objective. (Refer to MCWP 3-41.2, *Raids*.) Surprise, firepower, and violence of action are the keys to a successful raid. Patrols achieve surprise by attacking—

- When the enemy is least prepared (e.g., during periods of poor visibility such as darkness, rain, fog, or snow).
- From an unexpected direction. (This might be accomplished by approaching through a swamp or other seemingly impassable terrain.)
- With concentration of firepower at critical points within the objective.

Planning

A successful raid requires detailed planning. The leader of a combat patrol engaged in raiding must anticipate probable situations and decide upon definite courses of action to meet them. Rehearsals are imperative.

A raid patrol conducts such missions as destroying an enemy outpost or seizing prisoners from an observation post or lightly defended position.

While preparing for the mission, the patrol leader requests fire support required for the accomplishment of the mission. If practical, artillery and mortars should be employed to isolate the objective to prevent movement of enemy reinforcements into the area.

Execution

The leader's plan must be detailed and complete. All of the considerations outlined in chapter 11 must be covered. Patrol formations must provide for ease of control and all-around security while moving to and from the objective area and provide for rapid and coordinated deployment of the various units once the objective area is reached. The leader's plan usually includes the encirclement of the hostile position—either physically or by fire—in order to isolate it during the assault.

The final simultaneous assault against the objective develops when enemy defensive fires at the objective are suppressed by either friendly fire superiority or surprise. The assault is covered by the fire of the unit assigned the function of support by fire.

Grenades, SMAWs, and demolitions are most effective for clearing bunkers.

Security units are posted to isolate the objective. The patrol leader signals them when the withdrawal begins. As a minimum, security is on each flank and to the rear (at the objective rally point).

Actions at the Objective

The patrol leader halts the patrol near the objective at the final preparation position. Security is established and the leader's reconnaissance is made with appropriate subordinate leaders. When the leaders return to the patrol, they confirm previous plans or announce any changes. Movements are arranged so all units reach their positions simultaneously. This improves the patrol's capability for decisive action, if prematurely detected by the enemy.

The teams of the security element move to positions to secure the objective rally point, give early warning of enemy approach, block avenues of approach into—and prevent enemy escape from—the objective area. As the assault element moves into position, the security element informs the patrol leader of all enemy activity, firing only if detected or on the patrol leader's order. Once the assault element commences action, the security element prevents enemy entry into or escape from the objective area. The security element covers the withdrawal of the assault element (and support element, if employed) to the objective rally point, withdrawing only upon order or prearranged signal.

As the assault element approaches the objective, it deploys early enough to permit immediate assault if detected by the enemy. Each team uses stealth while moving into proper position. On command, or if one or more of the assault element is detected and fired upon by the enemy, the support element opens fire to neutralize the objective, then ceases or shifts fire according to prearranged plans and signals. As supporting fires cease or shift, the assault element assaults the objective. Demolition, search, and other teams are protected by the assault element while they work. On order, the assault element withdraws to the objective rally point.

If a support element is employed, its leader deploys teams to provide fire support for the assault element. Each member of the support element must know the scheme of maneuver to be used by the assault element, specific targets or areas to be neutralized by fire, and the signals that will be employed to commence, shift, and cease fires. The support element withdraws on order of the patrol leader. At the objective rally point, the patrol leader quickly reorganizes the patrol and begins the return movement to friendly lines.

13004. CONTACT PATROLS

General

Contact patrols establish and/or maintain contact to the front, flanks or rear by—

- Establishing contact with an enemy force when the definite location of the force is unknown.
- Maintaining contact with enemy forces through direct and/or indirect fires, or observation.
- Avoiding decisive engagement with the enemy.

Task Organization and Equipment

Task organization and equipment depend on the known enemy situation and anticipated enemy contact. A patrol sent out to establish contact with an enemy force is organized, armed, and equipped to overcome resistance of light screening forces in order to gain contact with the main enemy force. It is not organized and equipped to engage the main enemy forces in combat. Communication is paramount; radios must be reliable over the entire distance covered.

Actions at the Objective

The patrol leader selects a series of objectives. Once an objective is reached, the patrol leader initiates a planned set of actions in order to establish and maintain contact with the enemy. These plans and actions are guided by the missions to establish or maintain contact—not to engage in decisive combat. Contact with the enemy is maintained for the purposes of surveillance, applying pressure, and preventing seizure of the initiative. If the contact patrol becomes decisively engaged with the enemy, many of the tasks originally assigned to the patrol cannot be accomplished, since the enemy has seized the initiative and friendly forces are not forced to react.

13005. AMBUSH PATROLS

General

An ambush is a surprise attack from a concealed position upon a moving or temporarily halted target. It is one of the oldest and most effective types of military actions. Ambush patrols conduct ambushes of enemy patrols, resupply columns, and convoys. The intent of an ambush is to place the enemy in a dilemma where staying in the kill zone or attempting to move out of it prove equally lethal. The ambush may include an assault to close with and decisively engage the enemy, or the attack may be by fire only.

Purpose of Ambushes

Ambushes are executed for the general purpose of reducing the enemy's over-all combat effectiveness and for the specific purpose of destroying its units. The cumulative effect of many small ambushes on enemy units lowers enemy troop morale and harasses the enemy force as a whole.

Destruction is the primary purpose of an ambush because loss of men killed or captured, and loss of equipment and supplies destroyed or captured, reduces the overall combat effectiveness of the enemy.

Harassment, though less apparent than physical damage, is a secondary purpose of ambushes. Frequent ambushes force the enemy to divert men from other missions to guard convoys, troop movements, and carrying parties. When enemy patrols fail to accomplish their mission because they are ambushed, the enemy is deprived of valuable information. A series of successful ambushes causes the enemy to be less aggressive and more defensive minded. His men become apprehensive, overly cautious, reluctant to go on patrols, seek to avoid night operations, are more subject to confusion and panic if ambushed, and in general, decline in effectiveness.

Classification of Ambushes

A *deliberate ambush* is one in which prior information about the enemy permits detailed planning before the patrol departs for the ambush site. Information needed to plan a deliberate ambush includes the size, composition, and organization of the force to be ambushed; how the force operates; and the time it will pass certain points or areas. A deliberate ambush may be planned for such targets as—

- Any force if sufficient prior information is known.
- Enemy patrols that establish patterns by frequent use of the same routes or habitually depart and re-enter their own areas at the same point.
- Logistic columns.
- Troop movements.

An *ambush of opportunity* is conducted when available information does not permit detailed planning before the patrol departs. This is the type of ambush that an infantry unit normally conducts. An ambush of opportunity should not be confused with a hasty ambush. An ambush of opportunity is a planned ambush; a hasty ambush is an immediate action. In planning for an ambush of opportunity, the patrol must be prepared to execute any of several courses of action based on the types of targets that may be ambushed and must rehearse prior to departure. The course of action taken is determined when the opportunity for ambush arises.

The patrol leader may be directed to reconnoiter an area for a suitable ambush site, set up at the site selected, and execute an ambush against the first profitable target that appears.

The patrol may depart just after dark, move to a specific point, observe until a designated time, ambush the first profitable target after that time, and return before daylight.

A *hasty ambush* is an immediate action where the patrol makes visual contact with an enemy force and has time to establish an ambush without being detected. The actions for a hasty ambush must be well rehearsed and accomplished through the use of hand and arm signals given from the patrol leader.

Types of Ambushes

There are two types of ambushes: point and area. The *point ambush* is one where forces are deployed to attack along a single killing zone. The *area ambush* is one where forces are deployed as multiple related point ambushes.

A point ambush, whether independent or part of an area ambush, is positioned along the enemy's expected route of approach. Formation of the forces conducting the ambush is important because, to a great extent, the formation determines whether a point ambush is able to deliver the heavy volume of highly concentrated fire necessary to isolate, trap, and destroy the enemy.

The ambush formation to be used is determined by careful consideration of possible formations and the advantages and disadvantages of each in relation to—

- Terrain.
- Visibility.
- Forces.
- Weapons and equipment.
- Ease or difficulty of control.
- Target to be attacked.
- Combat situation.

For a detailed discussion of ambush formation, see appendix D.

Ambush Operation Terms

The *ambush site* is the location where an ambush is established.

The *killing zone* is that portion of an ambush site where fires are concentrated to trap, isolate, and destroy the target. On little-traveled roads, an obstacle placed in a defile, in the woods, on a bridge or on a steep upgrade can be used effectively to force vehicles to halt, and thus render the occupants vulnerable to attack. Antitank mines may be emplaced and the occupants of the wrecked vehicle killed or captured while still dazed by the explosion.

A *near ambush* is a point ambush where the attacking force is located within reasonable assaulting distance of the killing zone (50 meters is a guide). A near ambush is most often conducted in close terrain, such as a jungle or heavy woods.

A *far ambush* is a point ambush where the attack force is located beyond reasonable assaulting distance of the killing zone (beyond 50 meters is a guide). A far ambush may be more appropriate in open terrain offering good fields of fire or when the target will be attacked by fire only.

Factors for a Successful Ambush Patrol

There are many factors that give the ambush its best chance of success. The ideal situation would be to position the ambush on favorable terrain and have detailed planning completed beforehand.

The patrol leader of an ambush looks for the most favorable terrain in which the enemy is canalized between two obstacles with limited opportunity to attack or escape. Suitable areas include defiles, small clearings, bends in trails, and steep grades. Dense undergrowth adjacent to the ambush site permits observation from concealed positions. The ambush patrol should have maximum cover and concealment, not only for the firing positions, but for the routes of withdrawal. The enemy should be in an area offering as little protection from fire as possible. Favorable fields of fire include stretches of road, trail or open ground of at least 100 meters for machine guns and 15 meters for rifle fire and grenades. The ambush site can be improved by constructing obstacles—such as felled trees, wire, land mines, or booby traps—to impede the enemy.

Planning

A deliberate ambush or an ambush of opportunity requires thorough planning.

A deliberate ambush plan is based on extensive knowledge of the enemy and terrain, and is planned and rehearsed in great detail. A physical reconnaissance of the ambush site is made during the preparation phase and information gained is incorporated into the plan. All likely immediate actions of the enemy when ambushed are examined. Planned counteractions are developed and rehearsed.

In planning an ambush of opportunity, any available information on the enemy and terrain is used. A tentative plan for the ambush that incorporates all anticipated actions is developed and rehearsed. However, the bulk of planning is done concurrently during the patrol leader's reconnaissance of the prospective ambush site. In a rapidly developing situation, hasty ambush immediate action is employed.

The route and ambush site considerations apply to both deliberate ambushes and ambushes of opportunity. A primary route that allows the patrol to enter the ambush site from the rear is planned. Entering the prospective killing zone is avoided. If the killing zone must be entered to place mines or explosives, care is taken to remove any tracks and signs that might alert the enemy and compromise the ambush. If mines or explosives are to be placed on the far side of the ambush site, or if the appearance of the site from the enemy's viewpoint is to be checked, a wide detour is made around the killing zone. Care is taken to remove any tracks that might reveal the ambush. An alternate route from the ambush site to the objective rally point, as in other patrols, is planned.

Maps and aerial photographs are used to carefully analyze the terrain. When possible, an on-the-ground reconnaissance of the ambush site is made prior to occupation. Obvious ambush sites are avoided as the element of surprise is even more difficult to achieve in these areas. An ambush site must provide for—

- Favorable fields of fire.
- Occupation and preparation of concealed positions.
- Canalization of the target into the killing zone. (An ideal killing zone restricts the enemy on all sides, confining him to an area where he can be quickly and completely destroyed. Natural obstacles, such as cliffs, streams, embankments, or steep grades, are used whenever possible to force vehicles to slow down. Man-made obstacles, such as barbed wire, mines, and craters in the roads, are used to supplement natural obstacles.)
- Covered routes of withdrawal that enable the ambush force to break contact.
- Avoidance of enemy pursuit by fire.

Occupation of Ambush Site

The surrounding area is searched for enemy patrols prior to occupation of the ambush site. Ambush formations are used to physically deploy the patrol in a manner to inflict maximum destruction upon the enemy and to provide maximum security to the patrol. Ambush formations are contained in appendix D.

Positions

The patrol is moved into the ambush site from the objective rally point. Security is positioned first to prevent surprise while the ambush is being established. Automatic weapons are then positioned so each can fire along the entire killing zone. If this is not possible, overlapping sectors of fire are provided to cover the entire killing zone. The patrol leader then selects his position, located so he can tell when to initiate the ambush. Riflemen and grenadiers are positioned and sectors of fire are assigned to cover any dead space left by the automatic weapons. The patrol leader sets a time by which positions are to be prepared. Patrol members clear fields of fire and prepare positions in that order, with attention to camouflage for both.

Suitable Objective Rally Point

An easily located objective rally point is selected and made known to all patrol members. The objective rally point is located far enough from the ambush site so that it will not be overrun if the enemy assaults the ambush. Routes of withdrawal to the objective rally point are reconnoitered. Situation permitting, each man walks the route he is to use and picks out check-points. When the ambush is to be executed at night, each man must be able to follow his route in the dark. After the ambush has been executed, and the search of the killing zone completed, the patrol is withdrawn quickly but quietly, on signal, to the objective rally point where it reorganizes for the return march. If the ambush was not successful and the patrol is pursued, withdrawal may be by bounds. The last group may arm mines, previously placed along the withdrawal route, to further delay pursuit.

Local Security

Security must be maintained. Security elements do not usually participate in the initial attack, but protect the rear and flanks, and cover the withdrawal.

Patience

The Marines of the ambush force must control themselves so that the ambush is not compromised. Patience and self-discipline are exercised by re-maining still and quiet while waiting for the target to appear, particularly if the patrol occupies the ambush site well ahead of the arrival of the enemy. Patience is necessary so as not to alert the enemy to the presence of the ambush.

Surprise

Surprise must be achieved, or the attack is not an ambush. If complete surprise cannot be achieved, it must be so nearly complete that the target is not aware of the ambush until too late for effective reaction. Surprise is achieved by careful planning, preparation, and execution so that targets are attacked when, where, and in a way for which they are least prepared.

Coordinated Fires

Properly timed and delivered fires contribute heavily to the achievement of surprise, as well as to de-struction of the enemy. The lifting or shifting of fires must be equally precise; otherwise, the assault is delayed and the enemy has an opportunity to recover and react. All weapons, mines, and demolitions are positioned and all fires, including those of available artillery and mortars, are coordinated to achieve the following results:

- Isolation of the killing zone to prevent the enemy's escape or reinforcement.
- Surprise delivery of a large volume of highly con-centrated fires into the killing zone.

Control

Close control of the patrol is maintained during move-ment to, occupation of, and withdrawal from the ambush site. This is best achieved through rehearsals and establishment and maintenance of good com-munications. When the enemy approaches, the temp-tation to open fire before the signal is given is resisted. The patrol leader must effectively control all elements of the ambush force. Control is most critical at the time the enemy approaches the killing zone. Control measures must provide for—

- Early warning of enemy approach.
- Fire control. Withhold fire until the enemy has moved into the killing zone, then open fire at the proper time.
- Initiation of appropriate action, if the ambush is prematurely detected. Individual patrol members must be prepared to react if detected by the enemy prior to the initiation of the ambush.
- Timely and orderly withdrawal of the ambush force from the ambush site and movement to the objec-tive rally point.

It is important to remember that an ambush patrol should have four distinct signals: one to open fire (with an alternate signal to open fire to be used at the same time as the primary); a signal to cease fire or shift fire; a signal to assault or search the killing zone; and a signal to withdraw. The signal to open fire should meet two criteria: first, it should be the firing of a weapon that will kill the enemy; secondly, it should be a weapon reliable in any weather condition. A good primary signal is a Claymore mine, and an alternate signal would be a closed bolt weapon (M16A2). Open bolt weapons (M240G, M249) should not be relied upon to initiate an ambush.

Execution of an Ambush

The manner in which the patrol executes an ambush depends primarily on whether the ambush's purpose is harassment or destruction. To a lesser degree, the execution of the ambush is determined by whether the ambush is deliberate or an ambush of opportunity.

When the primary purpose is harassment, the patrol seals off the area with security teams to prevent enemy reinforcement and escape. Maximum damage is inflicted with demolitions and automatic weapons fire. The patrol delivers a very heavy volume of fire for a short time and withdraws quickly and quietly. The patrol avoids being seen by the enemy.

When the primary purpose of the patrol is destruction, the area is sealed off with security units. Maximum damage is inflicted with demolitions, antitank weapons, and automatic weapons fire from the support team or element. When these fires cease or shift, an assault is launched into the killing zone with heavy fire and violence to complete destruction. The assault unit provides security, while designated teams search and/or capture personnel and destroy vehicles and equipment. On the patrol leader's command, or by prearranged signal, all units withdraw to the objective rally point and move out quickly.

When the patrol's primary purpose is to obtain supplies or capture equipment, security units seal off the area. Demolitions and weapons are used to disable vehicles. The assault unit must use care to ensure its fire does not damage the desired supplies or equipment. Designated teams secure the desired items; other teams then destroy enemy vehicles and equipment.

The most successful ambush is one where the attacker is deployed and concealed in such a way that the enemy will unknowingly be surrounded by fire. The usual method is for the attackers to deploy themselves along a trail or route the enemy will travel. The enemy is permitted to pass by the center of the attacker's force so that the attack can be made from the front, flank, and/or rear. One or two men are posted well forward and to the rear along the route to prevent any enemy from escaping. All fires should be delivered simultaneously on a prearranged signal.

An effective method of luring the enemy is for an ambushing patrol to cut communication or electrical wire. The patrol then deploys and ambushes the line repair crew when it arrives. Since the line crew may be protected by riflemen, the attackers must be careful to engage the entire party.

Vehicles and foot personnel moving on well-established transportation routes can sometimes be captured by altering or moving directional signs so as to divert the enemy into an area where he can be more readily attacked. The attack can best be accomplished at an obstacle, such as a stream or gully, that forces the enemy to stop or slow down.

After the enemy has been ambushed and destroyed, the unit quickly withdraws over a prearranged route to the objective rally point. Speed is very important, since the noise of the ambush could alert other nearby enemy units.

13006. SECURITY PATROLS

General

Security patrols are assigned missions that may or may not require them to engage the enemy. They are used in proximity to defensive positions, on the flanks of advancing units or in rear areas. Purposes of security patrols are to detect infiltration by the enemy, destroy infiltrators, and protect against surprise and ambush.

In any situation where there is a threat of attack, such as a rear area threatened by guerrillas or a facility that is under threat of a terrorist attack, all Marines, not just the infantry, must know how to conduct a security patrol.

In just the offensive operations, infantry units provide security patrols to screen their flanks, areas, and routes. Whereas, in defensive operations, security patrols are used to prevent the enemy from infiltrating an area, detect and destroy infiltrators, and prevent surprise attacks. In rear areas, particularly when there is guerrilla or terrorist threat, the requirement to conduct security patrols increases for all Marine air-ground task force (MAGTF) units ashore, particularly aviation and combat service support units.

Task Organization and Equipment

Generally, a Marine rifle squad or similar sized organization is considered ideal for security patrols. Communications are important to higher headquarters so that they receive information from the patrol; and communications are important to the patrol to request fire support, etc. The radio the patrol carries must have the range necessary for higher headquarters to be able to receive transmissions from anywhere along the patrol route, and the patrol must have a secondary means of reporting (i.e., flare signals upon contact).

Patrol Procedures

All of the procedures presented in previous sections are to be used in security patrols.

Patrol Planning

Security patrol planning includes—

- Rehearsing prior to departing friendly lines.
- Maintaining communications.
- Support by organic weapons.
- Reinforced if necessary.
- Using varied routes and never establishing a routine pattern.
- Staying within proximity of friendly units.

Patrol Techniques

Within rear areas, an irregular pattern of patrol is established and changed daily. Outside of friendly lines it would be prudent to establish a definite preplanned route for the patrol, of which all adjacent units know the route. The parent unit commander establishes frequent checkpoints for control. If checkpoints are designated, the patrol leader treats them as individual objectives to be searched and cleared.

The patrol has a definite plan as to what to do if contact with enemy is made, how to break contact, how to defend itself, and how to call for supporting fires. It is imperative that patrol members know what to do if they become split or separated; i.e., location of rally points and how to be recovered.

13007. URBAN PATROLS

General

As national strategy continues to focus on regional vice global conflicts, the Marine Corps will continue to conduct urban operations in various operational environments. Cities and towns are often the center of economic and political power and are therefore extremely vulnerable to urban insurgent activities and violence. The lessons learned from recent operations in Somalia, as well as experiences gained by British forces in Northern Ireland, provide the foundations of urban patrolling.

Enemy Forces

Enemy forces in urbanized areas range from organized military forces to low intensity engagements with insurgents, such as terrorists or local gangs. For Marine Corps doctrine and tactics, techniques, and procedures (TTP) for dealing with the higher intensity threat in urbanized areas, refer to MCWP 3-35.3, *Military Operations on Urbanized Terrain*.

Common Threat Tactics

Cities provide cover and concealment for both friendly forces and enemy forces. However, enemy forces generally find active support only in certain areas of the town or city. The urban-based insurgent or terrorist usually lives in a friendly community or in one where the people are too frightened to withhold support or inform anyone about the situation. The insurgent or terrorist normally maintains close contact with leaders and others friendly to the cause. The enemy will often have efficient communication and intelligence systems, sometimes involving women and children to provide cover for its activities.

The urban insurgent normally cannot, like his rural counterpart, establish bases and recruit large military units. He is generally an individual or a member of a relatively small group. He relies on the cover afforded by the city's populace and terrorizing them to coerce loyalty or support.

Urbanized areas tend to give the insurgent and/or terrorist many opportunities to initiate action and gain advantage. The normal presence of large numbers of people in cities provides the foe an opportunity to mass crowds quickly and manipulate demonstrations easily. The presence of women and children during mass demonstrations may restrict the courses of action available to friendly forces. Major incidents stemming from overreaction or excessive use of force by friendly forces may provide the insurgent with propaganda material. Publicity is easily gained in an urbanized area because major incidents can't be completely concealed from the local population. Insurgent successes can be exploited to discredit the ability of host nation police, friendly forces, and the civil government, and gain recognition for the insurgents' cause.

The urban insurgent or terrorist can usually be expected to operate more boldly than his rural counterpart. This is reflected in the enemy's tactics. A single sniper or bomber may be the norm in the urbanized area, whereas the rural threat is generally the more conventional ambush. In urbanized areas, explosive devices can be easily emplaced and used effectively against large groups or select individuals. Enemy forces may be expected to employ the following tactics in urbanized areas:

- Using local communications, such as radio and newspapers, for propaganda purposes.
- Disrupting industry and public services through strikes and sabotage.
- Generating widespread disturbances designed to stretch the resources of the security force.
- Creating incidents or massing crowds in order to lure the patrol or reaction force into a trap.
- Provoking security forces in the hope that they may react improperly, therefore discrediting the security force by means of propaganda.
- Sniping at roadblocks, outposts, sentries, and patrols.
- Attacking friendly bases with rockets and mortars.
- Planting explosive devices, either against specific targets or indiscriminately, to cause confusion and destruction, and lower public morale and confidence.
- Using ambush patrols.
- Firing on friendly helicopters.

Principles of Urban Patrolling

Patrolling in an urban environment often presents conditions considerably different and often more complex than those encountered in rural and less inhabited areas. While the principles of patrolling are still relevant in an urban situation, the nature of urban patrolling has led to the development of six specific urban patrolling principles. They are—

1. **Depth.** The restrictive, canalizing nature of urbanized terrain usually limits a patrol's ability to disperse laterally. To prevent the patrol from bunching up, patrols normally maintain dispersion along the length of a patrol formation.

2. **Mutual Support.** The positioning of units in-depth within the patrol enables one unit to cover another unit's movement and facilitates immediate action during various situations. Aircraft, vehicles, and snipers also provide good mutual support.

3. **Deception and Pattern Avoidance.** Deception and pattern avoidance are normally a planning consideration of the headquarters directing the entire patrolling plan. By varying patrol routes, durations, and departure times, hostile actions commonly used against urban patrols—such as enemy ambushes, and roadblocks—are more difficult to plan and may be preempted.

4. **Intra-patrol Communication.** Elements of an urban patrol must have the means to communicate with each other. Ideally, each element will possess a radio enabling it to remain in continuous communication that facilitates rapid response and reporting to higher headquarters, reaction force coordination, and coordination of actions with other patrols or fire support agencies.

5. **Establishment of a Reaction Force.** The nature of urbanized terrain (its compartmentalization) makes urban patrols more vulnerable to a wider range of hostile actions. The requirement for immediate, coordinated reinforcement of a patrol is best satisfied by employment of an established reaction force. The reaction force requires superior mobility (relative to the enemy's) and fire support to be effective for this task.

6. **Three-Dimensional Threat.** Patrolling in an urban environment requires constant attention to its three-dimensional aspect; hostile actions can originate from rooftops, streets, subsurface levels or combinations of all levels at once.

Classification of Urban Patrols

Mission

The vast majority of urban patrols are overt in nature, with their presence readily apparent to the local populace. Most urban patrols are combat vice reconnaissance patrols. The vast majority of urban patrols are security patrols. However, units may be assigned secondary tasks of reconnoitering specific or general areas along the patrol routes.

Raids normally involve a swift penetration of an objective to secure information, confuse the enemy or destroy installations. Raids include a planned withdrawal upon completion of the assigned mission.

Movement

Means of movement are as follows:

- Dismounted: movement on foot.
- Vehicular: movement by motorized, mechanized or armored vehicle.
- Helicopterborne: movement by helicopter, however, helicopterborne patrols will usually involve dismounted or vehicular movement after patrol insertion.
- Combination: movement using a combination of methods.

Task-Organization

Units task-organize to meet the specific requirements of the mission and situation. The Marine rifle squad is ideally suited for urban patrolling and can easily integrate attached specialists required by specific missions. Specialists who may accompany urban patrols include—

- Interrogator-translator team (ITT) and counterintelligence team (CIT) Marines.
- Explosive ordnance disposal (EOD) personnel.
- Members of host nation or allied military forces.
- Interpreters.
- Local community leaders.
- Local law enforcement officers.
- Public affairs personnel escorting media representatives.

Dismounted Patrol Organization

Patrols should maintain unit integrity (fire team and/or squad) in organizing elements for an urban patrol. Similar to reconnaissance patrols, the vulnerability of urban patrols necessitates that all elements must provide for their own security in addition to the combat aspect of the mission. Combined assault and security teams are an effective method to organize for all-around security.

Urban Patrol Base Operations

Urban patrols may operate from an established patrol base that may be located within the unit's assigned area of operations or an area designated for a patrol. The patrol base should be located in a building used exclusively for this purpose. Patrol bases may also be located within a larger site that houses other agencies, such as a higher headquarters' command echelon.

If located within a larger site, the patrol base will be included in the overall facility security plan. If isolated, the patrol base must consider the following security factors:

- External security:
  - Barrier plan.
  - Sentry posts.
  - Local security patrols.
- Internal security:
  - Covered positions for all Marines.
  - Contingency plan for hostile actions against patrol base.

The Reaction Force

The high probability of an urban patrol becoming involved in a hostile or volatile incident requires the establishment of a dedicated reaction force for rapid reinforcement, support or extraction of the patrol. Ideally, the reaction force is—

- Large enough and task-organized in a manner that it can meet and quickly defeat the expected threat. Reaction forces are normally tiered with a lead element (normally one-third of the unit's size) and a main body force (the other remaining two-thirds of the force). Reaction force response times routinely are determined in advance by higher headquarters.
- Ready to respond immediately.
- Motorized or mechanized and supported by close air support and other fire support.
- Familiar with the area of operations.
- Briefed on the patrol's plans and monitors the status of patrols in progress.
- Task-organized to be multimission capable.
- Able to communicate with the higher headquarters, fire support assets, patrol base, and the patrol.
- Controlled by higher headquarters, once employed.

Patrol Preparation

Planning

Higher headquarters will—

- Designate the area for patrol.
- Provide intelligence briefs and updates.
- Ensure liaison with allied forces and the civilian populace.
- Provide special equipment and personnel required for the mission (scout snipers, public affairs officer, interpreters, etc.)
- Provide urban maps, photos, terrain models as required.
- Consider deception and pattern avoidance when issuing mission.
- Prescribe rules of engagement (ROE).

Intelligence Brief

An intelligence brief is conducted by the S-2 officer or representative prior to a patrol conducting its mission. The brief adresses the situation relevant to the specific patrol (e.g., routes, areas, updated enemy situations).

Coordination

Higher headquarters will effect liaison with adjacent and allied forces, as well as civil authorities and other agencies, having a possible effect on the patrol. The patrolling unit generally follows the same procedures as those used during patrol planning and execution in a jungle or forest environment.

The Urban Patrol Order or Warning Order

The Urban Patrol or Warning Orders use the same format and considerations as noted in this publication for patrolling. They rely heavily on a detailed terrain model, photographs, and subterranean construction to ensure complete understanding of the plan.

Rehearsals

The limited size of the patrol base usually precludes the need for full-scale rehearsals. Immediate action drills, such as crossing danger areas, are rehearsed in as much detail as possible, despite the limited available space.

Inspections

Initial and final inspections are conducted in the same manner as other patrols. Attached personnel must be fully integrated into the patrol and familiar with the plan and unit standing operating procedures (SOPs).

Conducting an Urban Patrol

Movement

Individual and unit movement considerations are generally the same as those for other patrols. However, urban environments require consideration of additional factors. Because of these factors, an urban patrol leader should—

- Ensure that each movement within a patrol takes place under the observation or cover of another individual or element of the patrol.
- Know where cover can be taken in the event of a hostile incident or action.
- Be prepared for contact with civilians, especially children, during the patrol and be aware that they may intentionally attempt to distract patrol members.
- Expect the presence of vehicles (both moving and stationary) along the patrol route.
- Expect members of the patrol to be approached by dogs and what action to take if threatened.

Patrol Formations

Squad-sized Patrols. The need for immediate fire power outweighs the dangers of becoming canalized. In contrast to other types of patrols, the headquarters element of an urban patrol will normally locate at the lead of the patrol column. This allows the patrol leader greater flexibility through control of two combined assault and security (A&S) teams. The leaders of these elements tactically stagger their members on each side of the road (see fig. 13-1 on page 13-12).

A&S teams follow in trace of the headquarters element and maintain unit integrity on separate sides of the street. One unit will remain slightly to the rear to create a staggered interval between Marines on either side of the street. This allows A&S teams to take lateral routes in support of headquarters element without having to cross a street to do so.

Platoon-sized Patrols. Squads will generally travel abreast of each other, moving along parallel routes. The interval between squad-sized units and/or teams is situation-dependent, but is usually between 100 and 150 meters (roughly two city blocks; this often prevents visual contact between the units). The intent is to create less of a target to an aggressor, yet still allow the patrol to quickly react to an incident. Individuals within units or teams will move in a staggered column as in a squad-sized patrol. See figure 13-2, which is located on page 13-13).

Night

Night patrols will generally be at least squad-sized and will generally use the same formation as that for day patrols. At night, it may be necessary to close distances between individuals or elements to maintain control. Consideration should be given to the use of night vision devices and thermal weapons sights.

Navigation, Control, and Security Measures

The patrol leader is ultimately responsible for the navigation. The headquarters element normally functions as the base unit during movement. The designated navigator is normally assigned from within the headquarters element. City maps are often inaccurate or outdated; however, when used with aerial photographs and other navigational aids, they can be effective for urban navigation.

Checkpoints and phase lines should be related to major streets (alleys, buildings, bridges) for easy identification. Arrival at checkpoints and crossing of phase

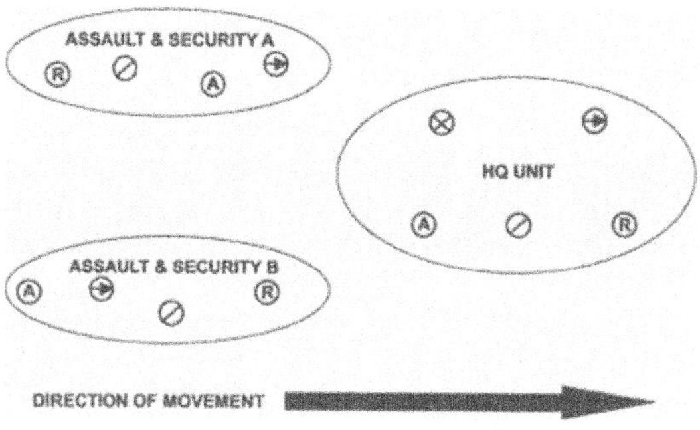

Figure 13-1. Squad-sized Dismounted Urban Patrol Formation.

lines should be relayed to higher headquarters using established brevity or codewords. A detailed patrol overlay is issued to both the reaction force and higher headquarters to keep them advised of routes and control measures used. A terrain model should be maintained at higher headquarters to aid in supervision and control of the patrol.

The use of camouflage should be limited to avoid frightening and confusing the local populace. The often overt nature of urban patrols may negate the need for camouflage. Patrols are sometimes deployed to show force presence and usually move on the urban street in plain view.

During daylight, patrols will routinely vary their rate of movement ranging from short halts to brief periods of double-timing. The British term for this urban patrolling technique is *hard-targeting*, meaning it makes the patrol harder for an enemy to target. Altering the rate of movement is intended to frustrate the enemy's ability to coordinate an attack or ambush against a targeted patrol.

Patrols should use short security halts, with Marines taking up mutually supporting firing positions. Marines must always work in pairs, ensuring mutual support. The last Marine in the element will provide rear security, but stays in his buddy's sight.

Individual Tasks

Individuals may be assigned collateral tasks performed throughout the patrol that may increase the patrol's situational awareness. Individual tasks may include—

- Vehicle spotter: looks for suspicious or known insurgent vehicles.
- Personnel spotter: observes and attempts to recognize previously identified enemy in crowds.
- Talker: attempts to gain information from casual conversations with the local populace. (Talkers are usually subordinate leaders or Marines with foreign language skills.)
- Searcher: conducts physical searches of vehicles and personnel while other patrol members provide cover and security.
- Marksman: engages point targets when the tactical situation does not permit massed or high volume fire. Other patrol members provide security to cover the marksman's engagement.

Departure of Friendly Lines

Urban patrols must vary their departure times to prevent being ambushed while exiting the patrol base.

Individual elements will usually depart exits at staggered times and at different movement rates, especially where sentries cannot provide cover. (This technique is used when the environment adjacent to the patrol base is dangerous.)

Once an element has exited friendly lines, a short halt is conducted in a predetermined, covered initial rally point 50 to 100 meters from the base. This ensures all elements are in position before the patrol continues.

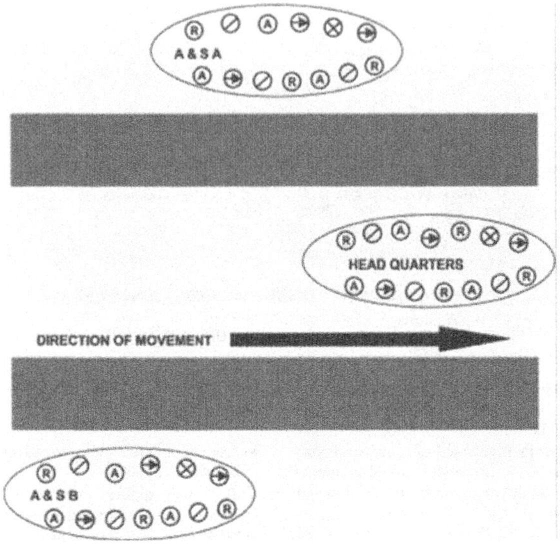

Figure 13-2. Platoon-sized Dismounted Urban Patrol Formations.

Exit points, routes from the base, departure techniques, and locations of IRPs should be varied constantly to avoid setting a pattern. This pattern avoidance may also include using vehicles to insert patrols away from the urban patrol base and employing empty vehicles as part of a deception plan.

Danger Areas

Urban patrols may encounter hundreds of danger areas during a single patrol. The three-dimensional threat requires keen situation awareness by every patrol member. Many danger areas can be dealt with simply by avoidance, while others require an adjustment of patrol formation, movement rate, etc. In the urban environment, places to be treated as danger areas are points that pose a major threat to the patrol, such as local political and religious headquarters, weapons containment areas, roads and routes that canalize movement and direct fire, and any area with a history of repeated contact.

Near and far side rally points are designated and briefed during the issuance of the patrol order. Squad-sized formations may use the A&S teams to provide flank security for the headquarters element and for each other. The headquarters element identifies the danger area and takes up a position on the near side of intersection. Individuals provide all-around security (see fig. 13-3 on page 13-14).

Two Marines (one from each A&S team) are designated to move through the headquarters element and establish respective firing positions on the near side of the danger area covering the patrol's near side flanks. They are followed by a second pair (again, one Marine from each A&S team) that moves across to the far side of the danger area and establishes respective firing positions covering the patrol's far side flanks (see fig. 13-4 on page 13-14).

Once near and far side flank security is established, the headquarters element moves across to the far side of danger area (see fig. 13-5 on page 13-15).

The remaining A&S team members then cross the danger area and join the headquarters element on the far side (see fig. 13-6 on page 13-15).

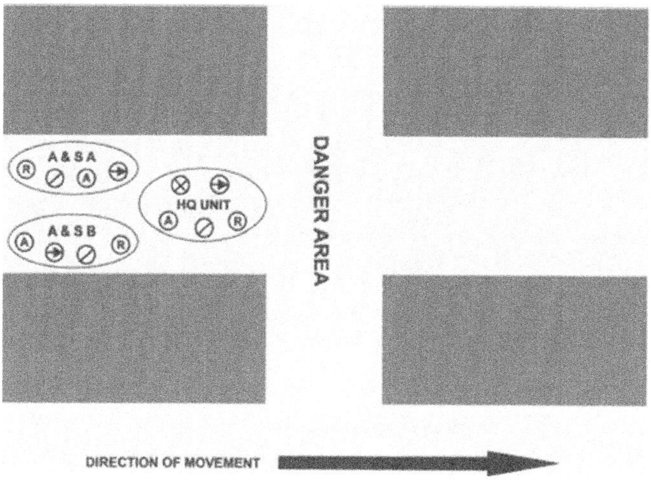

Figure 13-3. Approaching Danger Area.

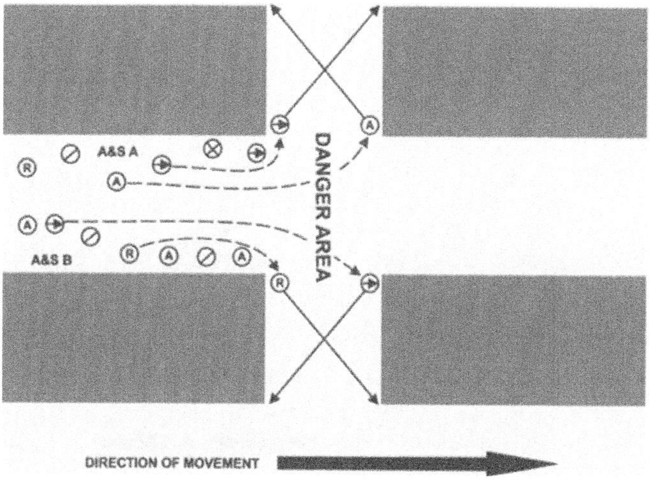

Figure 13-4. Securing Flanks.

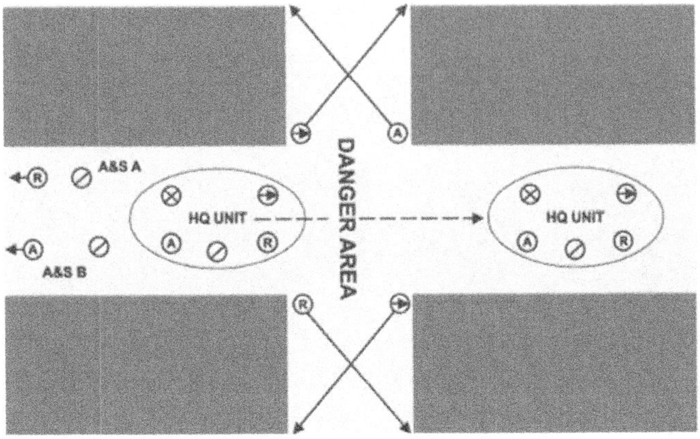

Figure 13-5. HQ Element Moves Through.

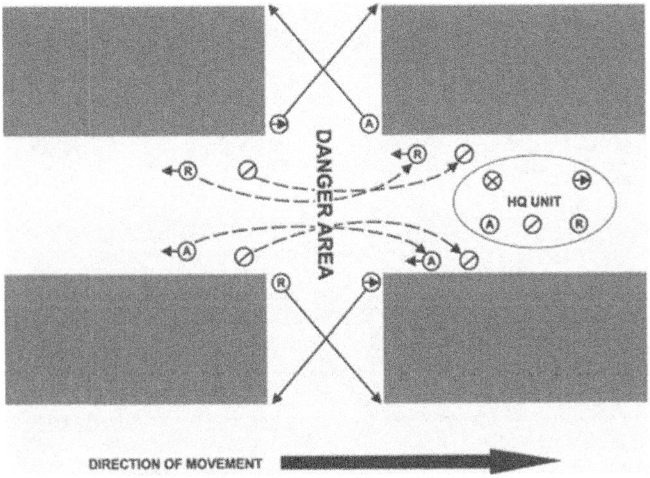

Figure 13-6. A&S Elements Move Through.

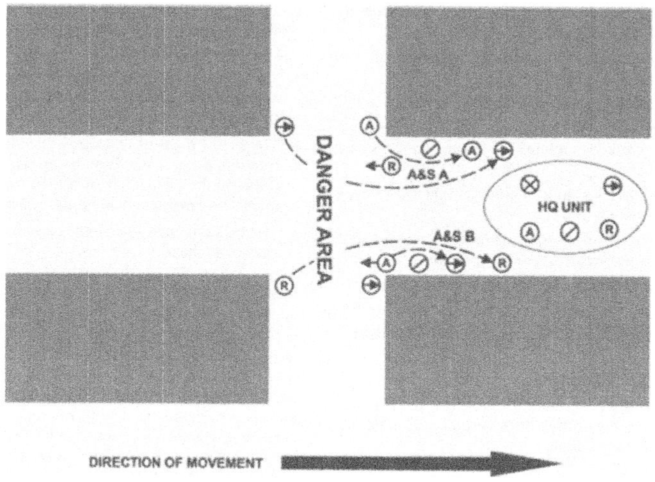

Figure 13-7. Patrol Resumes Movement Along Intended Route.

Once the trail A&S team members cross the danger area, they take up rear security and cover the movements of the near and far side flank security teams as they return to their positions in the patrol formation. The near side security team should collapse back first, followed by their far side counterparts. The patrol then resumes its advance away from the danger area (see fig. 13-7).

Interaction with Local Populace

Urban patrols must interact with the local populace. Patrols are at first a novelty to the civilians but can quickly become an unwelcome intrusion. The movement of the patrol must be fast enough to prevent the enemy from massing their fires upon it, but deliberate enough to ensure adequate security and mutual support. Patrol members must realize that they are usually the only Marines the local populace will encounter and that an inappropriate gesture, comment or act could lead to the deterioration of rapport between U.S. forces and the general population. Marines must remember that the vast majority of the individuals with whom they come in contact will be noncombatants attempting to survive in trying political, economic, and social situations.

Hostile incidents often seem to occur spontaneously, but there are usually indications that can alert Marines to imminent danger. The most obvious are the sudden alteration of normal routines, patterns, and attitudes of the local populace or other unusual activity. Some examples include—

- Observers on rooftops, in windows, etc., who are obviously tracking the patrol.
- The unusual absence of pedestrian traffic and people on porches.
- Stores, markets or street vendors closed suddenly or without explanation.
- Changes in civilian attitude toward patrol members.
- Unknown individuals or vehicles in the patrol area.
- Unfamiliar vehicles parked in the patrol area (possible car bomb).
- Roadblocks.
- Children throwing rocks at patrols to possibly draw the patrol's attention away from a more serious danger, such as a deliberate ambush.
- Vehicles riding unusually low due to overloading (possibly ferrying people, weapons, explosives).
- Agitators trying to provoke an incident with patrol members.

- Absence of the usual stray dogs (dogs are adept at sensing danger and avoiding it).
- Anti-American graffiti suddenly appearing in the patrol area.
- Pictures of enemy leaders and martyrs posted in the patrol area.
- Civilian workers failing to appear at U.S. or friendly bases.
- Normal deliveries and pick-ups conducted late or early without reason.
- Sudden change of civilian sentiment in newspaper articles, radio broadcasts or other media.
- Women and children leaving to live elsewhere.

Immediate Actions Upon Enemy Contact

Reaction to Sniping

Snipings are often executed from a single firing point, but coordinated snipings delievered from multiple points are not uncommon.

Patrol element or team leaders should constantly try to identify likely firing points and anticipate their own reactions to a possible shooting. Normally, snipers in an urban environment have a detailed withdrawal plan. Once a patrol comes into contact with a sniper, the patrol leader must immediately assess the situation and maneuver his patrol accordingly. The patrol's mission, location, size, ROE, and location of the threat often determine whether the patrol will attempt to neutralize the targeted sniper. If the patrol leader decides to kill or capture the sniper, he uses planned and rehearsed immediate actions to maneuver and counter the sniper's assault. The goal is to kill the sniper or cut off his escape and capture him. There are three immediate reactions to neutralizing a sniper: initial contact, immediate follow-up, and subsequent follow-up.

Initial Contact. The initial contact is made when the sniper fires the first shot. The patrol must react immediately and positively to get behind the firing position in order to kill or capture the gunman. The period of contact ends when the gunman is killed or captured, or the patrol element or team leader on the scene ends it. The following technique is the same for both squad- and platoon-sized patrols:

- The element or team in contact attempts to identify the firing position and maneuvers designated marksman into position to return well-aimed and controlled fire. Other members of the patrol take up positions to cover the marksman's engagement.
- The patrol element or team leader in contact sends initial contact report to the patrol leader, who notifies higher headquarters.
- The element or team leader in contact determines appropriate cut-off positions and relays them to flanking elements or teams.
- The patrol element or team leader in contact continues to observe the firing point, but does not enter it due to the possibility of booby traps. Flank elements or teams set up along likely escape routes.
- The incident ends when either the sniper ceases fire or is neutralized.

Immediate Follow-Up. Regardless of the fate of the gunman, isolation of the firing point is necessary to prevent reinforcement and preserve forensic evidence (scent, spent casings, etc.). If not under fire, members of the patrol element or team cordon off the area surrounding the firing point. Flanking elements or teams maintain their positions and prevent civilians from entering the area. The patrol leader moves to link-up with the element or team in contact (if not his own), and makes an estimate of the situation. The patrol leader sends a SPOTREP to higher headquarters.

Subsequent Follow-Up. The aim of the subsequent follow-up is to use follow-on forces to clear the building of remaining resistance or to obtain evidence that can be used to capture the gunman. The patrol leader establishes a position where he can brief arriving units (reaction force commander, S-2 representative, EOD personnel, etc.). Once the arriving units have been briefed, recommendations are made to higher headquarters via radio. No one is allowed into the cordon without the patrol leader's approval.

Reaction to Becoming Decisively Engaged

If a patrol becomes decisively engaged from numerous firing positions, the following immediate action should be taken:

- All patrol members move to available cover and return accurate fire on identified firing points.
- The patrol leader assesses the situation and makes a decision to either request the reaction force or break contact.
- If the reaction force is requested, the patrol will maintain its position until the reaction force arrives. The patrol should use fire and maneuver to gain better tactical positioning and support the arrival of the reaction force. When the reaction force arrives,

its commander may decide to either clear occupied buildings or cover the patrol during its extract.

Reaction to Bomb Threat or Discovery

The use of command-detonated explosive devices is a common ambush tactic employed by a terrorist or insurgent in an urban environment. The appropriate response to a reported threat or an actual discovery is generally involves four steps (known as the four Cs)—

1. The patrol leader **CONFIRMS** the presence of the suspicious item.

2. Without touching or moving anything suspicious, patrol elements or teams **CLEAR** the immediate danger area to a minimum of 100 meters. The area is cleared from the suspected device outward, inform civilians as to the reason for evacuation.

3. A&S teams establish a **CORDON** to secure the cleared area. Avenues of approach are cordoned off to keep people out and to protect EOD or engineer personnel clearing the device. The assistant patrol leader acts as the cordon commander and informs the patrol leader when the cordon is secure. An effective cordon technique is to tape off the area with engineer tape, creating both a physical and psychological boundary.

4. **CONTROL** of the area is maintained throughout the bomb clearing operation by the patrol leader. The patrol leader sends a report to a superior concerning details of the device (if known) and the area affected. The patrol leader coordinates with arriving personnel (EOD, engineers, etc.). The patrol leader maintains communication with the assistant patrol leader and keeps the Marines informed of the progress of the clearing operation.

Reaction to a Bomb Detonation

Bombs may be used by an insurgent as a means of initiating an ambush on mounted or dismounted patrols, in which case the actions for decisive engagement apply. Immediate action in response to an isolated explosion is similar to that used in reaction to a sniping and breaks down into the same three phases:

Initial Contact. The patrol leader attempts to identify the likely initiation point and sends an initial contact report to higher headquarters. If the bomb was command-detonated, the patrol leader sends his A&S teams deep to cut off the bombers' escape routes. Any casualties are moved a minimum of 100 meters from explosion and out of the line of sight to it.

Immediate Follow-Up. The A&S teams may need to maneuver to positions behind the likely initiation point to cut off escape. Once in position, personnel checks are conducted and any suspects are detained. The patrol leader coordinates requests for required support (MEDEVAC, reaction force, etc.)

Subsequent Follow-Up. Due to the possibility of secondary detonations, the four Cs (confirm, clear, cordon, and control) can be conducted as in reaction to a bomb discovery or bomb threat.

Civil Disturbances

Urban patrols must prepare to react to spontaneous aggression by the local populace. In many cases, civil disturbances are organized by the enemy to draw dismounted patrols into a targeted area, or to distract them from enemy activity occurring elsewhere. Civil disturbances are generally divided into two categories: minor aggressive actions, and full-scale rioting.

Minor aggressive actions are activities characterized by rock-throwing or use of devices such as Molotov cocktails and may either be directed at the patrol or take place between different ethnic factions of the population. Minor aggressive actions are normally spontaneous in nature and may have minimal or limited objectives for the insurgents.

Full-scale rioting events are usually in response to another major event or incident that may enflame the populace. Full-scale riots are well-planned and orchestrated, with clear objectives or targets in mind. At times, patrols will need to attempt to maintain control of a civil disturbance situation; however, dismounted and mobile small unit patrols should generally avoid potential flashpoints. Procedures to handle civil disturbances are as follows:

- The patrol leader reports the incident to headquarters and attempts to diffuse the crowd by talking to crowd leaders.
- If the patrol leader determines the size of the disturbance is too large for the force to handle, the patrol should move away from the disturbance to a safer, more remote covered area and occupy positions to observe and report the situation to higher headquarters. To prevent the patrol from being pursued by the crowd, the patrol should move quickly and change direction, often at road junctions, to gain distance from it.

Patrol members should maintain dispersion to create a more difficult target. They should face the crowd at all times to see and avoid any projectiles thrown. Individual self-discipline must be maintained throughout the disturbance. Marines charging into the crowd or throwing objects back at the crowd will only worsen the situation. If pursued or trapped, the patrol leader may consider using riot control agents (combat support, pepper gas, etc.) to disperse or slow a crowd's movement. The use of riot control agents must be authorized under the established ROE. If the patrol leader believes the situation is deteriorating beyond the patrol's ability to control it, the patrol leader should request the reaction force, which may be better equipped to handle a large riot or mob.

Break Contact

As with patrols in rural areas, the patrol leader may be forced to break contact as a result of decisive engagement with the enemy. On the basis of his estimate of the situation, the patrol leader will normally break contact in one of the following ways:

- As a patrol, with elements providing cover for movement as defined by clock direction and distance.
- As individual units/teams taking separate routes out of the area, then linking up at a designated rally point a safe distance away from the engagement.

As in any contact with enemy forces, smoke may be employed to screen movement. Fire support agencies can be utilized to suppress targets; riot control agents can be employed to disrupt enemy movement.

Reentry of Friendly Lines

The reentry of a dismounted patrol into an urban patrol base is no different from that of a patrol conducted in a rural area. The same planning considerations and control methods apply.

Missions Related to Urban Patrolling

House Calls

House calls missions are usually part of a coordinated effort to collect information within the area of operations. They involve obtaining up-to-date information on particular houses and occupants. When possible, local police should accompany patrols to do the actual talking to the occupants. If this is not possible, a technique that may be used by an urban patrol is—

- A&S elements move to provide cover around the target house.
- The headquarters element provides security just outside the house.
- The patrol leader and one Marine for security enter the house, if invited. If not invited, they talk to the occupants on the doorstep. Patience and tact are required in requesting information. An interpreter should be present when language differences exist.

Vehicle Checkpoints

Urban terrorists or insurgents commonly use vehicles to transport personnel, weapons, explosives, and equipment. Civilian vehicles are often used for these purposes, creating the requirement to check as many vehicles as possible. While permanent, fortified checkpoints may be conducted along approaches into an urbanized area, dismounted patrols can be employed to establish hasty vehicle checkpoints to stop vehicles and to keep the enemy off-guard. The two common types of vehicle checkpoints patrols establish are hasty and deliberate.

Hasty checkpoints are deployed anywhere based upon the decision of the patrol leader. Patrols must not set patterns through the frequent use of the same sites.

Deliberate checkpoints are tasked by higher headquarters to achieve a specific purpose. Time and locations are carefully considered to avoid setting patterns.

The general layout for a squad-sized, two-way dismounted checkpoint is depicted in figure 13-8 on page 13-20. The technique is as follows:

- The A&S teams are positioned stealthily in blocking positions on both sides of the road.
- Both the patrol leader and assistant patrol leader act as "talkers" for each direction of traffic (with local police or interrogator or translator Marines acting as interpreters) while a two-man team from the headquarters element physically searches the vehicles.
- Obstacles or parked vehicles may be employed to create a staggered roadblock in center of the checkpoint to slow approaching vehicles.
- The checkpoint location should be sited so that approaching vehicles cannot see it until they have passed a security team, and they have no escape route then available.
- Signs announcing the checkpoint should be displayed a safe distance from the search area for safety to both drivers and Marines.
- Normally higher headquarters will issue criteria that determines which vehicles are searched, but random checks of cars should normally be made as

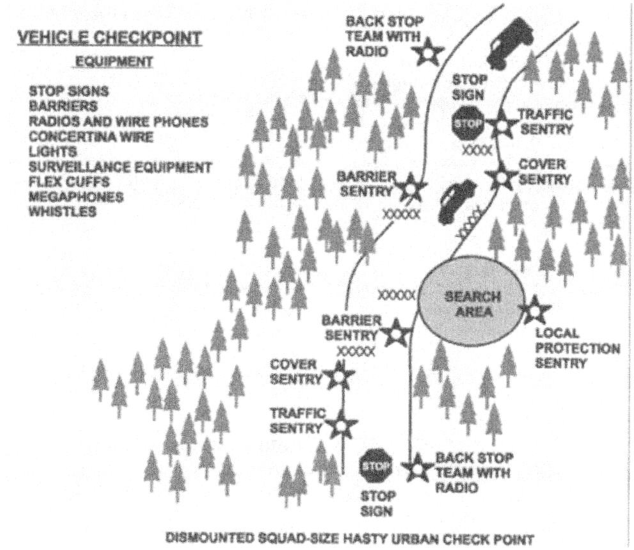

Figure 13-8. Dismounted Squad-sized Hasty Urban Vehicle Checkpoint.

well. While the vehicle is being searched the driver should accompany the searcher around the vehicle.

- Vehicle occupants should be made to exit the vehicle and then searched. Whenever possible, women should be searched by female police or female Marines.
- All patrol members should conduct themselves with courtesy and politeness. If nothing is found, an apology for the inconvenience is recommended.
- A hasty vehicle checkpoint should not be conducted any longer than 30 minutes for security reasons.
- The ROE should dictate whether or not action should be taken against vehicles that fail to stop at the checkpoint. Failure of a vehicle to stop does not automatically give authorization to fire.

Observation Posts

Urban observation posts are established to provide extended security, not only for patrol bases but also for patrols operating within the observation post's sector of observation. Observation posts can be established in conjunction with sniper operations and for providing overwatch for patrols operating within their sector of observation. Observation posts are nor-mally positioned on dominating terrain or in buildings outside the patrol base itself.

Insertion to the observation posts and conduct of observation may be either overt or clandestine in nature. Overt observation posts usually will be hardened positions to increase security. A patrol provides cover while the observation post is being inserted. A clandestine observation post relies on stealth of insertion and occupation for protection. It is normally positioned in abandoned buildings to cover sectors of observation that overt observation posts cannot. Because of their nature, clandestine observation posts are difficult to successfully establish and should not be manned for an extended period of time. Orders establishing observation posts (and patrols) must address the method of extraction as well as actions upon compromise/attack.

Cordon and Search

The cordon and search mission involves isolating a predesignated area by cordoning it off and systematically searching for enemy personnel, weapons, supplies, explosives or communications equipment. While large-scale cordon and search operations are planned and rehearsed in advance and normally entail extensive coordination with local law enforcement agencies, a squad-sized urban patrol may often conduct a cordon and search of a point target—searching one house or building identified by intelligence as a possible weapons cache.

The basic principle of a search of a populated area is to conduct it with limited inconvenience to the population. The populace may be inconvenienced to the point where they will discourage urban guerillas or insurgents from remaining in the area, but not to the point that they will assist the enemy as a result of the search.

Upon receiving intelligence that warrants the searching of a building or a specific tasking from higher headquarters, A&S elements of the patrol move to establish an inner cordon around the target building to seal it off, with the primary intent of preventing movement out of the targeted building.

On order, the designated reaction force deploys to establish an outer cordon, oriented outward some distance from the inner cordon and covering routes leading into the area in order to prevent outside interference/reinforcement. The reaction force maintains a reserve to reinforce either cordon or react to unfolding events (civil disturbance in response to the operation).

Once the cordons have been established, the patrol leader, with the assistance of local police or interpreters, informs the local populace that a building is about to be searched, that a house curfew is in effect (if permitted by higher headquarters), and that all occupants should remain indoors. Occupants of the target house are instructed to gather at a central location to stay out of the way of the search party.

The headquarters element, having linked up with any required assistance (explosive ordnance disposal (EOD), ITT, etc.) now acts as the search party and accompanies local police. A female searcher should be included in the party, if necessary.

Occupants are searched and screened first for possible enemy personnel. Apprehended persons are evacuated as soon as possible.

The head of the household should accompany the search party throughout the operation to be able to counter incriminating evidence and possible accusations of theft and looting against Marines. If possible, a prominent member of the local community should act as a witness.

Buildings are best searched from top to bottom. Ideally, the search is conducted with the assistance of combat engineers using mine detectors to locate hidden arms and ammunition.

If the targeted building is empty or the occupant refuses entry, it may be necessary to forcefully enter the premises to conduct a search of the dwelling. If an unoccupied house containing property is searched, arrangements should be made with the local community to secure it until its occupants return. Unnecessary force and damage to property should be avoided during the search.

Motorized Urban Patrols

The advantages of a motorized urban patrols is their ability to capitalize on the speed, mobility, and protection offered by various vehicles. They may be motorized, mechanized or armored vehicles or a combination. Generally, motorized urban patrols possess greater combat power than dismounted patrols and can cover larger areas faster than dismounted patrols.

The disadvantages of motorized urban patrols is that they are restricted to roads and are vulnerable to ambush by the enemy. They are also restricted in their ability to interact with the local populace.

Motorized patrols are generally organized in the same manner as dismounted patrols (see fig. 13-9). Unit integrity is maintained when assigning personnel to specific vehicles.

The urban patrolling principles apply to motorized patrols in much the same manner as dismounted patrols. Mutual support and depth are achieved by maintaining constant observation between vehicles and coordinating support with any dismounted patrols in the area. All-around security is achieved through the use of constant observation as well as the vehicle's mobility and firepower. Positive communications between units or teams are maintained through vehicle radios. Patrol routes and speeds are varied to promote deception or pattern avoidance.

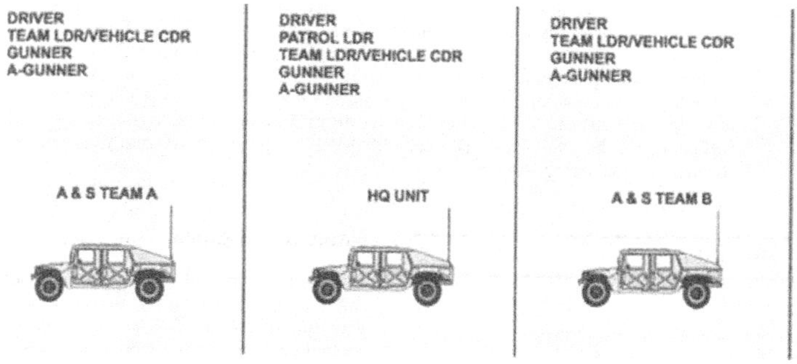

Figure 13-9. Organization of a Squad-sized Motorized Patrol.

The canalizing nature of streets and alleys force vehicular patrols to use a traveling overwatch movement technique to reduce vulnerability to ambushes. All vehicles travel at a moderate rate of speed with the lead vehicle stopping only to investigate potential danger areas. If vehicles must stop in danger areas, designated crew members' will dismount to provide security. The gunner will remain at the ready and in the turret while the driver remains in the driver's seat with the vehicle running.

Vehicles should move at a high rate of speed only when responding to an incident. At all other times, vehicle speed should be between 15 to 20 mph to allow for quick reaction and good observation. Distances between vehicles should be approximately 50 meters (one half to one city block) or such that visual contact and mutual support are ensured. Particular care is taken at major road junctions and other danger areas to ensure individual vehicles do not become isolated.

Vehicles with doors removed generally enhance observation and overall security, yet expose Marines to thrown objects, theft and concealment.

Chapter 14. Information and Reports

It is necessary that patrol leaders and all patrol members be trained in observing and reporting their observations accurately. The leader of a patrol should have all members of the patrol immediately signal or report any information obtained. These reports should not be restricted to information about the enemy, but should also include information about the terrain, such as newly discovered roads, trails, swamps, and streams. The leader includes all information in the report to the officer dispatching the patrol. (Refer to MCRP 2-15.3, *Reconnaissance Reports Guide,* for detailed report formats.)

14001. REPORTING

The officer dispatching the patrol instructs the patrol leader on whether and when messages are to be sent back during the patrol and what communication means to use. Messages may be oral or written. They must be accurate, clear, and complete. Every message should answer the question what, where, and when. For a detailed discussion on reporting, see paragraph 6002.

Verbal Messages

A patrol leader sending a verbal message should make it simple, brief, and avoid using numbers and names. The messenger should accurately repeat the message back to the patrol leader before leaving.

Written Messages

In preparing written messages, the patrol leader must distinguish between fact and opinion. Information about the enemy should include: strength; armament and equipment; actions; location and direction of movement; unit destination, if known; time enemy was observed; and the patrol's location when the observation was made. Use of an overlay or sketch may often simplify the message.

Messengers

A messenger team is given exact instructions as to where to deliver the message and the route to take. Any information obtained along the route should be reported at the time the message is delivered. Messengers must be given all practical assistance. If in danger of capture, the messenger immediately destroys the message.

Radio and Signals

If the patrol is provided with a radio, a definite radio schedule for checking in must be arranged before departure of the patrol. The patrol leader takes every precaution to ensure that codes and copies of messages are not captured by the enemy. If a close reconnaissance of enemy lines is required, the radio should be left in a concealed location at a safe distance from the enemy. Once a report is sent by radio, the patrol should immediately leave the area to avoid the possibility of detection by enemy locating devices. Pyrotechnics (flares, colored smoke, grenades) and air-ground panels may also be used by patrols for reporting information by a prearranged signal.

SALUTE Report

Information must be reported as quickly, accurately, and as completely as possible. An established method to remember how and what to report about the enemy is to use the acronym SALUTE:

Size
Activity
Location
Unit
Time
Equipment

An example of such a report is: "Seven enemy soldiers, unit unknown, traveling SW, crossed road junction on BLACK RIDGE at 211300 August carrying one machine gun and one rocket launcher."

14002. CAPTURED ITEMS

Every patrol should make a practice of searching enemy casualties, prisoners, and installations first for booby traps, then for equipment, papers, maps, messages, orders, diaries, and codes. Search techniques should be practiced often in order to reduce the time exposed to potentially dangerous situations. Items found are collected by the patrol leader and turned in with the patrol report. The items found are marked as to time and place of capture. When possible, captured items should be linked to a specific prisoner who possessed the items or to the place where the items were found. When this is done, the enemy prisoner of war tag and item tag are marked accordingly. The patrol leader must impress upon the members of the patrol the importance of turning in all documents and equipment. Furthermore, the patrol leader must ensure all information gathered by the patrol is quickly disseminated.

14003. PRISONERS

A patrol normally does not capture prisoners unless required by the mission. If prisoners are taken, the "5S and T" rule applies. This memory aid stands for: search, segregate, silence, speed, safeguard, and tag.

Search. Prisoners are body-searched thoroughly for weapons and documents as soon as they have been captured. This search must include the helmet, body armor, and gas mask. These items are left with the prisoner for protection until the patrol is completed. Weapons, equipment, and documents are tagged and immediately sent to the patrol leader.

Segregate. Prisoners are segregated into isolated groups: officers, noncommissioned officers (NCOs), privates, deserters, and civilians. By segregating prisoners, it makes it more difficult for leaders to organize escapes and issue orders to subordinates.

Silence. Silence is essential. Do not allow prisoners to talk to each other.

Speed. Speed is required in getting prisoners to the commander who dispatched the patrol. Timely information secured from prisoners is essential.

Safeguard. Prisoners are safeguarded as they are moved. They are restrained, but not abused. If the patrol will soon reach friendly positions, prisoners are not given cigarettes, food or water until they have been questioned by interrogators. If the patrol will not return to a friendly position for a lengthy time, the prisoners are given food, water and medical aid.

Tag. Equipment and personal effects recovered from prisoners are tagged so that they may later be matched back to specific individuals.

14004. PATROL REPORT

Every patrol leader makes a report when the patrol returns. Unless otherwise directed, the report is made to the person ordering the patrol. If the situation permits, the report is written and supported by overlays and/or sketches. The patrol leader's report should be a complete account of everything of military importance observed or encountered by the patrol while on the assigned mission. It should include the following information:

- Size and composition of patrol.
- Tasks and purpose (mission).
- Time of departure.
- Time of return.
- Routes, out and back (show by sketch, azimuth, trace on map).
- Terrain (general description to include any man-made or natural obstacles and critical terrain features which, if occupied by either enemy or friendly forces, would allow them to control the surrounding area).
- Enemy (size, activity, location, unit, time, equipment).
- Any map corrections (show on map).
- Miscellaneous information not covered elsewhere in report.
- Results of enemy encounters.
- Condition of patrol, including disposition of any dead or wounded.
- Conclusion and recommendations.

14005. PATROL CRITIQUE

After the patrol has rested and eaten, the patrol leader should hold a critique. Constructive criticism is made. It is an excellent time to prepare for future patrols by going over lessons learned as a result of the patrol.

Appendix A. Patrol Warning Order

The warning order is issued as soon as practical with all available information included to assist patrol members in preparation.

1. Situation. Friendly and enemy situation information necessary for initial preparation.

2. Mission. Statement of what the patrol is to accomplish, and the purpose for accomplishing it. When, how, and where will be discussed in the patrol order.

3. Execution

a. Task Organization. General patrol organization and assignment of responsibilities if known; otherwise, promulgate in the patrol order.

b. Tasks. Alert subordinate leaders to patrol tasks (e.g., stream crossing, helicopter rappelling, demolitions) requiring preparation prior to departure.

c. Coordinating Instructions

(1) Time schedule and location for individual preparation, rest, briefings, in-spection, rehearsal, and departure.

(2) Time, place, uniform, and equipment for receiving the patrol order.

(3) Tasks for subordinate leaders to direct and supervise the initial preparation which may include drawing ammunition, rations, and special equipment; conducting immediate action drills or other necessary individual or unit training; meeting and briefing attachment personnel; reconnoitering the area for passage of lines; and coordinating with the necessary unit leaders.

(4) Preliminary guidance to specialists and key individuals regarding their roles and organization within the patrol.

4. Administration and Logistics

a. Individual uniform, equipment, weapons, and prescribed load of rations, water, and ammunition.

b. Crew-served weapons (if required) and guidance regarding distribution of weapons and ammunition during movement.

c. Special equipment requirements (wire cutters, demolitions, radios, flashlights, infrared equipment, mines, binoculars) and their distribution during movement.

d. Restricted or prohibited items.

5. Command and Signal

a. Designation of assistant patrol leader and his role in preparation.

b. Designation of navigators and radio operators (if required).

c. Brief outline of patrol leader's schedule for preparation, and where he can be reached.

Appendix B. Patrol Order

A patrol order follows a warning order. Any of the following subjects that have been addressed in the warning order may be omitted with the exception of the mission statement. A patrol order is more detailed than a 5-paragraph squad order in that a great deal of attention is given to individual duties.

The patrol order also provides orientation information, both enemy and friendly, that impacts the patrol and enemy forces. This information can include, but is not limited to, weather, terrain, visibility, NBC considerations, local population situation, terrain model and/or map orientation, and behavior.

1. Situation

a. Enemy Forces

(1) Composition, disposition, and strength are based on size, activity, location, unit, time, and equipment (SALUTE).

(2) Capabilities and limitations to defend, reinforce, attack, withdraw, and delay (DRAW-D).

(3) Enemy's most probable course of action.

b. Friendly Forces

(1) Mission of next higher unit (task and commander's intent).

(2) Adjacent unit missions (task and intent). Also identify left, front, right, and rear.

c. Attachments and Detachments. (date and time effective).

2. Mission. The mission is the task to be accomplished, and its purpose (who, what, where, when, and why). For patrols, specify if the mission or time has priority.

3. Execution

a. Commander's Intent and Concept of Operations

(1) Commander's intent.

(2) The concept of operations tells the where, how, and who and lays out the patrol leader's general scheme of maneuver and fire support plan. It outlines the following:

(a) Task organization of the patrol.

(b) Movement to the objective area, to include navigation method.

(c) Actions in the objective area.

(d) The return movement, to include navigation method.

(e) Use of supporting forces (including illumination, if required).

b. Subordinate Element Missions. Subordinate element missions (task and purpose) are assigned to elements, teams, and individuals, as required.

c. Coordinating Instructions. This paragraph contains instructions common to two or more elements, coordinating details, and control measures applicable to the patrol as a whole. At a minimum, it includes—

(1) Time of assembly in the assembly area.

(2) Time of inspections and rehearsals (if not already conducted).

(3) Time of departure and estimated time of return.

(4) Location of departure and reentry of friendly lines and the actions associated with departure and reentry.

(5) Details on the primary and alternate routes to and from the objective area.

(6) Details on formations and order of movement.

(7) Rally points and actions at rally points.

(8) Final preparation position and actions at this position.

(9) Objective rally point and actions at this point.

(10) Actions at danger areas.

(11) Actions in the event of enemy contact.

(12) Details on actions in the objective area not covered elsewhere.

(13) Estimated time of patrol debriefing upon return.

4. Administration and Logistics

a. Changes/additions to uniform, equipment, and prescribed loads from that given in the warning order.

b. Instructions for handling wounded and prisoners.

5. Command and Signal

a. Command Relationships. Identify key leaders and chain of command.

b. Signal. Challenge and password, arm and hand signals, special signals, and radio frequencies and call signs.

Appendix C. Patrol Evaluation Checklist

This appendix provides a comprehensive checklist of critical patrolling steps, techniques, and procedures to aid unit leaders to critique patrol performance during training. Unit leaders may use the list as they observe the performance of a patrol to provide the patrol leader with a detailed analysis of the performance.

Warning Order	Yes	No
Did the PL ensure all patrol members were present before issuing the warning order?		
Did the PL issue a brief statement on the enemy situation?		
Did the PL issue a brief statement on the friendly situation?		
Did the PL state the mission in a clear, positive manner?		
Did the PL list all members of the patrol including attachments?		
Was the chain of command for the patrol covered fully?		
Were all members of the patrol asigned positions and duties in particular squads and teams?		
Were all the necessary individual duties assigned?		
Did the PL follow established principles in organizing the patrol into squads and teams?		
Was each patrol member assigned a particular weapon to carry on patrol?		
Did the PL list all special equipment needed to accomplish the mission?		
Was required special equipment assigned to the proper element to carry?		
Did the PL select uniform and equiment common to all based on METT-T?		
Coordination with Adjacent Units		
Did the PL coordinate with other patrols operating to the right and left?		
Was the route out and back coordinated?		
Was the time of departure and return cocordinated?		
Were call signs and frequencies coordinated?		
Was a signal for the FPF coordinated so as not to approach friendly lines during this time?		
Coordination With Front Line Units		
Did the coordinator pass the size of the patrol?		
Was the time of departure and return coordinated?		
Did the coordinator give a general area of oeprations for the patrol?		
Did the coordinator ask for information on known or suspected enemy positions and/or obstacles?		
Did the coordinator ask about information on the latest enemy activity?		
Did the coordinator ask for detailed information on friendly fire support available and the unit's barrier plan?		
Was the location of the IRP established and coordinated?		
Did the coordinator ask the forward unit to monitor their patrol frequency?		
Was the current challenge and password confirmed?		
Did the coordinator request that all information coordinated be passed on to any relieving unti?		
Patrol Order		
General		
Did the PL check to ensure all patrol members were present before issuing the patrol order?		

	Yes	No
Did the PL issue the patrol order in a forceful, confident manner?		
Did the PL make maximum use of available visual aids in issuing the patrol order (i.e., terrain model, sand table, map board, chalkboard)?		
Did the PL issue the patrol order in correct sequence?		
Did the PL issue the entire patrol order without allowing interruptions by patrol members?		
Did the PL adequately answer all questions asked by patrol members		
Did paragraph 1a (Enemy Situation) include—		
A weather forecast for the period of operation?		
A description of the terrain over which the patrol was to operate?		
Identification or description of enemy units known to bein the area of operations?		
Known locations of enemy units?		
Recent activity of enemy units?		
Stength of enemy units in the operating area?		
Did paragraph 1b (Friendly Situation) include—		
Mission of the next higher unit?		
Locations, missions, and planned actions of units on right and left?		
Fire support available to support the patrol (as per the coordination)?		
Missions and routes of other patrols operating in the immediate area?		
Did paragraph 1c (Commander's Intent) include—		
All views and ides of what the commander wants to be accomplished?		
The final result desired by the commander?		
Did paragraph 1c (Attachments and Detachments) include—		
All attachments to patrol and effective time of attachment?		
All detachments from patrol and effective time of detachment?		
Did paragraph 2 (Mission) include, at a minimum—		
Who was to conduct the patrol?		
What the patrol was tasked to do (i.e., conduct a point recon patrol)?		
Where the action was to take place?		
Why the action was to take place (i.e., the purpose)?		
Did paragraph 3 a (Commander's Intent and Concept of Operations) include—		
A complete concept of operations?		
A detailed description of the mission of elements?		
A detailed description of the mission of teams?		
A detailed description of the duties of specific individuals (i.e., navigator, compass man, pace m an, corpsman, APL)?		
Time of departure and time of return?		

	Yes	No
Type of formations and order of movement to be used?		
Description of the route and alternate routes(s) to include azimuths and distances between checkpoints as a minimum?		
Techniques to be used in the departure from friendly areas?		
Techniques to be used for the reentry into friendly areas?		
Location of the IRP and a tentative ORP, to include grid coordinates and recognizable terrain features?		
Method for designating and passing rally points?		
Actions to be taken at rally points if their use became necessary?		
Actions to be taken in the event of enemy contact?		
Actions to be taken at danger areas?		
Complete, detailed description of actions to be taken at the objective?		
Times and locations for rehearsals and the order of priority for rehearsals?		
Times and lcoations for inspections and the methods of conduct (to include uniform and equipment to be worn and/or carried)?		
Where, when, and by whom the debriefing is to be conducted?		
Did paragraph 4 (Administration and Logistics) include—		
Rations to be carried or reference to the warning order if there is no change?		
Arms and ammunition to be carried or reference to the warning order if there is no change?		
Uniform and equipment to be worn and carried or reference to the warning order if there is no change?		
Method for handling wounded or dead?		
Method for handling prisoners?		
Did paragraph 5a (Signal) include—		
Complete description of all signals, code words, frequencies, and call signs to be used within the patrol?		
Were signals adequate?		
Radio call signs to be used?		
Primary and alternate frequencies to be used?		
Call signs and frequencies to be used with other units in the area of oeprations (if applicable)?		
Required reports to be transmitted to higher headquarters, along with occasions for reporting?		
Code words and brevity codes to be used between the patrol and higher headquarters?		
Current challenge and password to be used in friendly held areas?		
Current challenge and password to be used forward of FEBA?		
Did paragraph 5b (Command) include—		
Chain of command or reference to the warning order if there is not change?		
Location of the PL during all stages of operation?		

	Yes	No
Location of the APL during all stages of operation?		
Inspection Rehearsal		
General		
Did the P L adequately use the allotted inspection time?		
Was the inspection conducted in an orderly manner?		
During the inspection, did the PL check the following:		
Completeness and correctness of uniform?		
Completeness of all equipment necessary to accomplish the mission?		
Operational condition of equipment?		
Did the PL question members to ensure they knew—		
The mission of the patrol?		
The concept of operations?		
Their individual duties and responsibilities?		
Chain of command and succession?		
Duties and responsibilities of key personnel?		
Did the PL utilize the rehearsal area to conform as much as possible to area of operations?		
Were the following major actions rehearsed:		
Actions at the objective?		
Actions at danger areas?		
Actions on enemy contacts?		
Departure and reentry of friendly lines?		
Did the PL critique each action after it had been rehearsed?		
Were interpatrol communications and control measures checked at rehearsal?		
Did the PL maintain control of the patrol during the rehearsal?		
Movement		
Did the PL use proper formations for movement?		
Did the PL adequately control the point team?		
Did the PL check the compass men?		
Was the PL aware of his position at all times?		
Was light and noise discipline enforced?		
Was the rate of movement appropriate for the mission?		
Did the PL make full use of pace?		
Did the PL make full use of count?		
Was proper security maintained during movement?		
Did the PL select and pass proper rallying points?		
Did the PL ensure that all members knew location of rallying points?		

	Yes	No
Did the PL properly use arm and hand signals?		
Did the PL recognize and halt the patrol a safe distance from a danger area?		
If the PL conducted a map check, did he conduct it properly (light discipline, security, etc.)		
Did designated individuals reconnoiter the far side of a danger area properly?		
Were support teams emplaced properly?		
Did the PL use a proper formation in crossing the danger area?		
Were reports made on enemy contact and at checkpoints?		
Actions on Enemy Contact		
Did the PL halt patrol a safe distance from a tentative ORP?		
Did the PL issue an adequate frag order before departing to look for a tentative ORP?		
Did the PL take appropriate personnel to look for the ORP?		
Did the PL select a suitable ORP?		
Did the PL secure the site adequately?		
Did the patrol occupy the ORP as stated in the patrol order or subsequent frag order?		
Did the patrol occupy the ORP in an orderly manner?		
Did the PL maintain control of the patrol during occupation of the ORP?		
Did the PL issue a satisfactory contingency plan to the APL before departing on a leader's recon?		
Did the PL take appropriate personnel on a leader's recon?		
Did the PL maintain or ensure that the patrol leader had communications with the patrol?		
Did the patrol avoid being detected by the enemy during the leader's recon?		
If contact was made, did the PL take appropriate action?		
Was security maintained during the leader's recon?		
Did the PL ensure the objective was kept under surveillance?		
Were appropriate orders given to surveillance teams?		
If a leader recon proved a tentative ORP unsuitable, did the PL move the ORP?		
Was a satisfactory leader's recon conducted?		
Was the ORP move conducted in a satisfactory and orderly manner?		
On return to the ORP, did the PL issue appropriate frag order and allow enough time for dissemination?		
Actions at Objective		
Did the PL issue a frag order for action at the objective?		
Was security emplaced prior to the teams departing from the ORP?		
Did the PL employ the terrain at the objective to the best advantage?		
Was stealth maintained while moving into the objective?		

	Yes	No
Did the Pl exercise positive control of squads, teams, and individuals at the objective?		
Were actions at the objective in accordance with the details outlined in the patrol order?		
Did the PL make use of supporting arms at the objective?		
Was the action at the objective successful?		
Was withdrawal from the objective accomplished quickly and orderly?		
Did units withdraw according to the details reflected in the patrol order?		
Was reorganization at the ORP completed in an expeditious and orderly manner?		
Did the patrol withdraw from the ORP quickly and quietly?		

Appendix D. Ambush Formations

Formations discussed here are identified by names which correspond to the general pattern formed on the ground by deployment of the assault element.

Line Formation

The assault element is deployed generally parallel to the route of movement of the enemy. This positions the assault element parallel to the long axis of the killing zone and subjects the target to heavy flanking fire. The size of the force which can be trapped in the killing zone is limited by the area the assault element can effectively cover. The enemy is trapped in the killing zone by natural obstacles, mines, demolitions, and direct fires. (See fig. D-l.)

A disadvantage of the line formation is the chance that lateral dispersion of the target may be too great for effective coverage. The line formation is appropriate in close terrain that restricts enemy maneuver, and in open terrain where one flank is protected by natural obstacles or can be protected by mines and demolitions. Similar obstacles can be placed between the attack force and the killing zone to provide protection from possible enemy counter attack. When an ambush is deployed in this manner, access lanes are left so that the enemy can be assaulted. (See fig. D-2.)

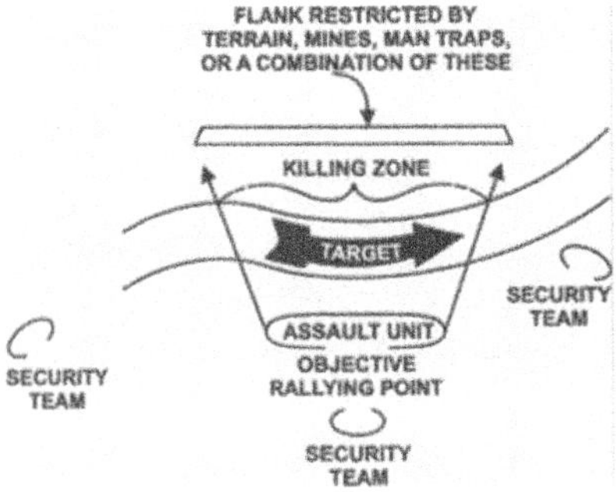

Figure D-1. Line Formation (Harassing or Destruction Ambush).

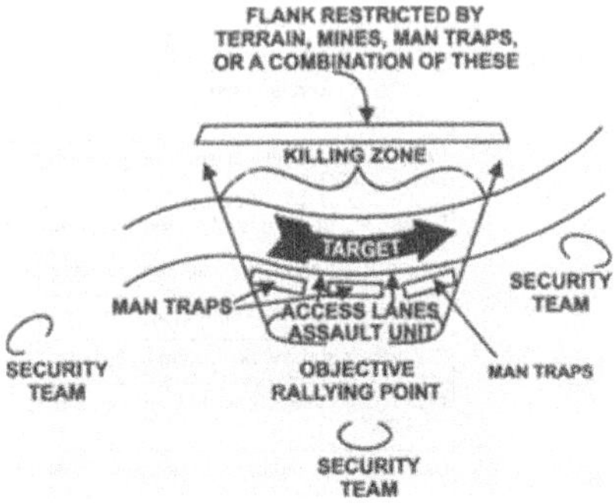

Figure D-2. Line Formation (Access Lanes for Assault of Target).

The main advantage of the line formation is its relative ease of control under all conditions of visibility.

L Formation

The "L" shaped formation is a variation of the line formation. This formation is very flexible because it can be established on a straight stretch of a trail or road (see fig. D-3) or at a sharp bend in a trail or a road (See fig. D-4 on page D-2).The long side of the assault element is parallel to the killing zone and delivers flanking fire. The short side of the attack force is at the

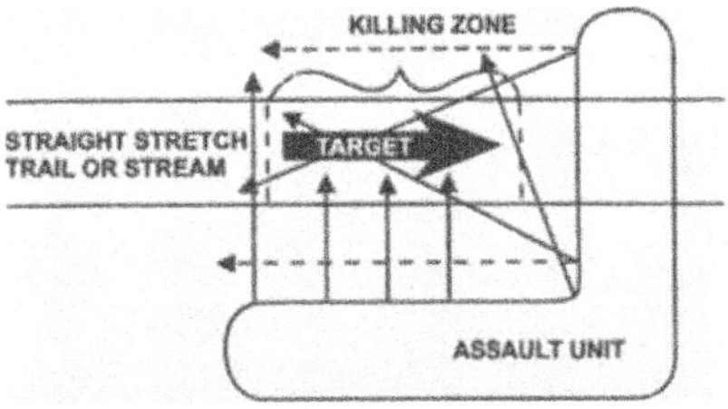

Figure D-3. "L" Formation (Destruction Ambush).

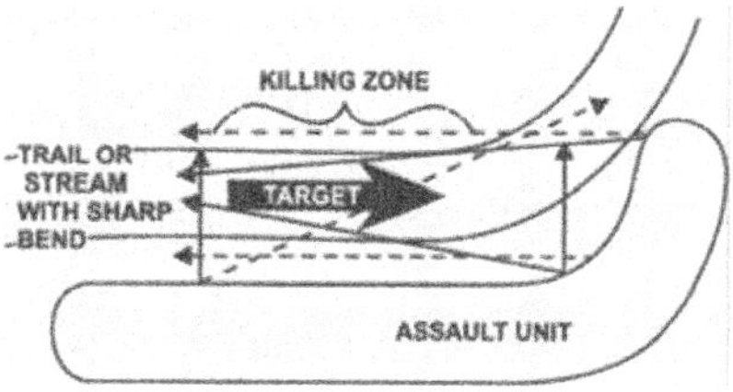

Figure D-4. "L" Formation (Bend of Trail or Stream).

end of, and at right angles to, the killing zone and delivers enfilading fire that interlocks with fire from the long side. When appropriate, fire from the short side can be shifted to parallel the long side if the enemy attempts to assault or escape in the opposite direction. In addition, the short side prevents escape and reinforcement. (See fig. D-5.)

"Z" Formation

The "Z" shaped formation is another variation of the line formation.

The assault element is deployed as in the "L" formation, but with an additional side so that the formation resembles a "Z". (See fig. D-6).

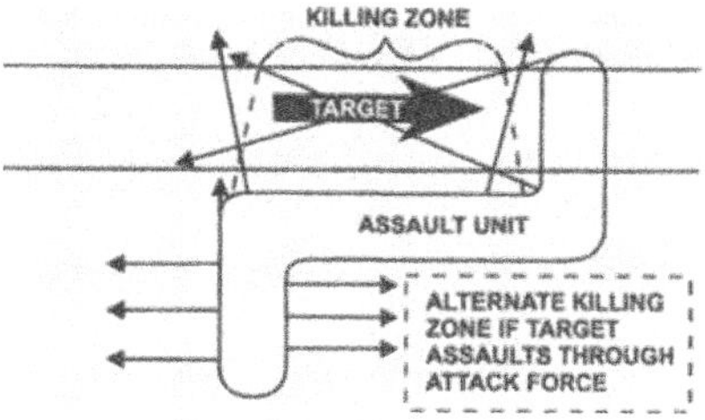

Figure D-6. "Z" Formation.

The additional wing may serve any of the following purposes:

- To engage an enemy force attempting to relieve or reinforce the enemy unit engaged in the kill zone.
- To seal the end of the killing zone.
- To restrict a flank.
- To prevent an envelopment.

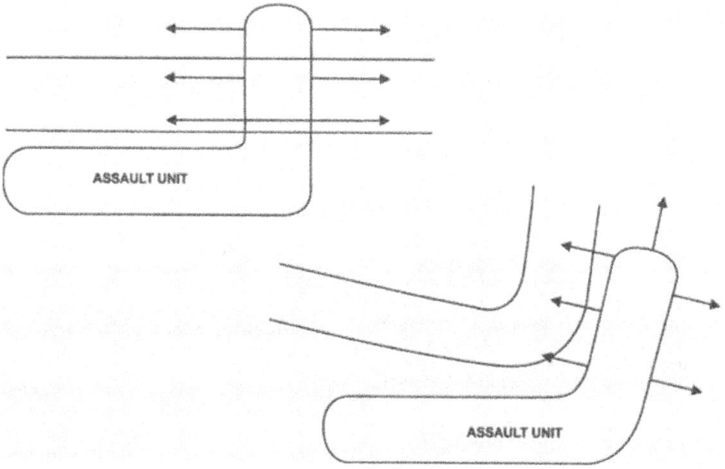

Figure D-5. "L" Formation.

"T" Formation

In the "T" shaped formation, the assault element is deployed across (perpendicular to) the enemy's route of movement so that its position forms the crossing of a "T" at the top. (See fig. D-7.)

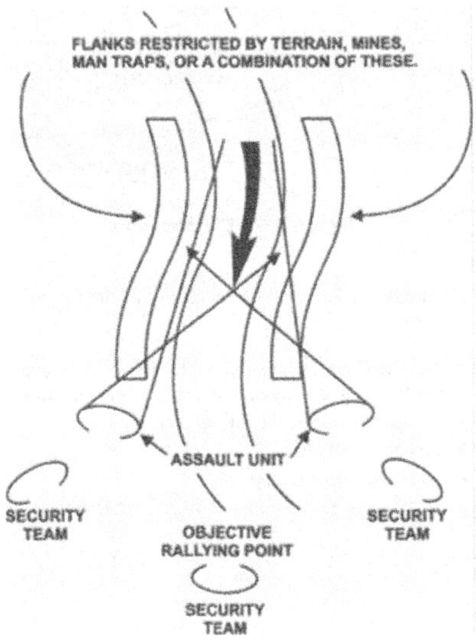

Figure D-7. "T" Formation.

This formation can be used day or night to establish an ambush to interdict movement through open areas that are hard to seal off.

A small force can use the "T" formation to harass, slow, and disorganize a larger force. When the lead elements of the enemy are engaged, they will normally attempt to maneuver right or left to close with the ambush. Mines and other obstacles placed to the flanks of the killing zones slow the enemy's movements and permit the ambush force to deliver heavy fire and withdraw without becoming decisively engaged.

The "T" formation can be used to interdict small groups attempting night movement across open areas. For example, the assault element is deployed along an avenue of approach with every second man facing the opposite direction. The attack of the enemy approaching from either direction requires only that every second man may shift to the opposite side of the formation. Each man fires only to his front and only when the enemy is at a very close range. Attack is by fire only and each man keeps the enemy under fire as long as it remains to his front.

If the enemy attempts to escape in either direction along the killing zone, each Marine takes the enemy under fire as the enemy comes into the Marine's sector of fire. The "T" formation is very effective at halting infiltration. But it has one chief disadvantage; there is a possibility that the ambush will engage a superior force at night while spread out. (See fig. D-8.)

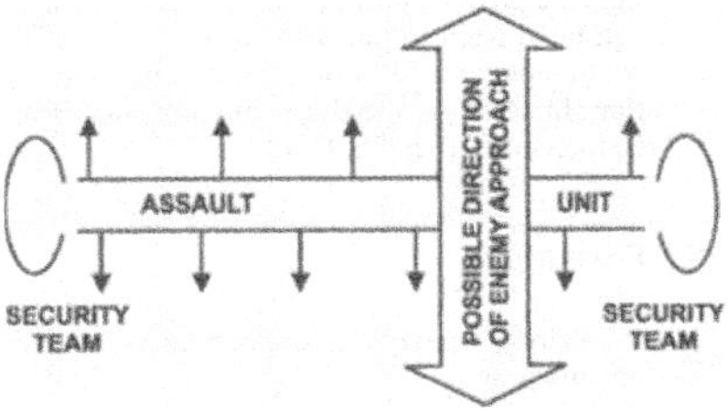

Figure D-8. "T" Formation (Target Approaching from Either Direction).

The "V" Formation

The "V" shaped formation is deployed along both sides of the enemy's route of movement so that it forms a "V"; care is taken to ensure that neither group (within the "V") fires into the other.

This formation subjects the enemy to both enfilading and interlocking fire. The "V" formation is best suited for fairly open terrain but can also be used in close terrain. When established in close terrain, the legs of the "V" close in as the lead element of the enemy force approaches the apex of the "V", and opens fire at a close range.

Here, even more than in open terrain, all movement and fire must be carefully coordinated and controlled to ensure that the fire of one wing does not endanger the other wing. The wider separation of forces makes this formation difficult to control, and there are few sites that favor its use. Its main advantage is that it is difficult for the enemy to detect the ambush until it is well into the killing zone. (See figs. D-9 and D-10.)

Triangle Formation

The triangle is a variation of the "V" formation and can be varied in three ways:

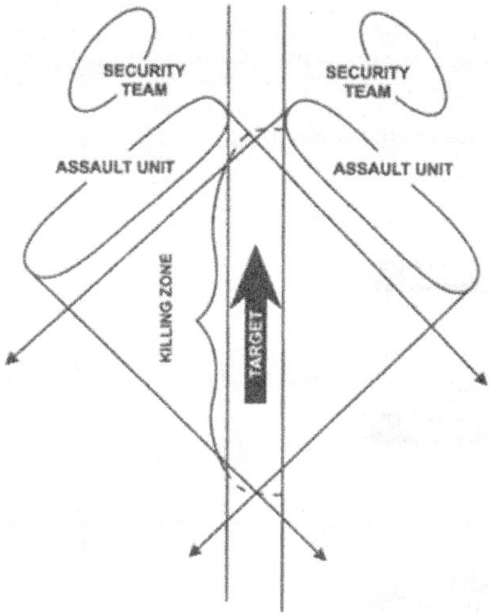

Figure D-9. "V" Formation (Open Terrain).

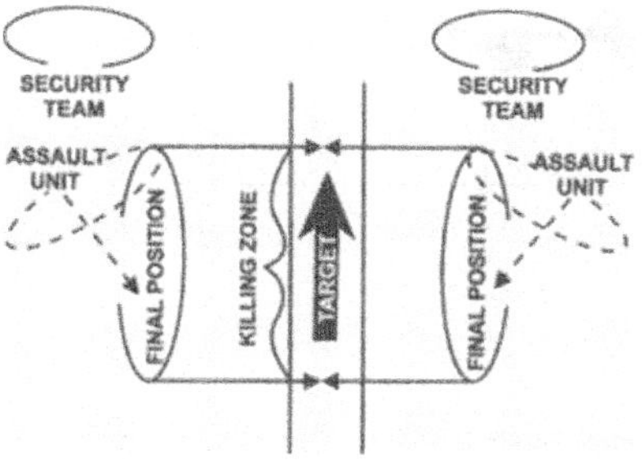

Figure D-10. "V" Formation (Close Terrain).

Closed Triangle Formation

(See fig. D-11.) The assault element is deployed in three teams, positioned so that they form a triangle (or closed "V"). An automatic weapon is placed at each point of the triangle and positioned so that it can be shifted quickly to interlock with either of the others. Men are positioned so that their sectors of fire overlap. Mortars may be positioned inside the triangle.

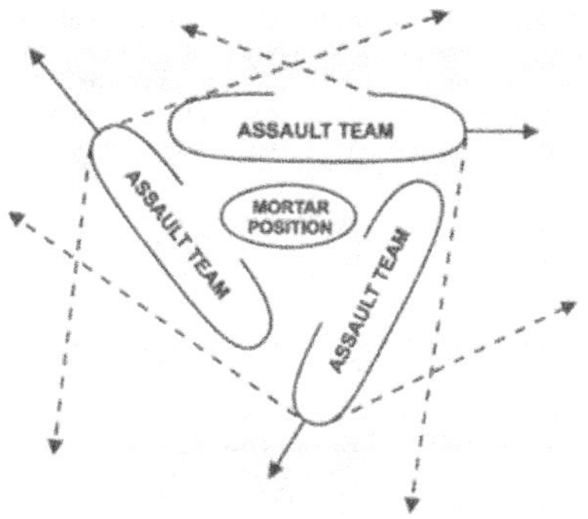

Figure D-11. Closed Triangle Formation (Night Ambush).

When deployed in this manner, the triangle ambush is used to interdict night movement through open areas. When enemy approach is likely to be from any direction, this formation provides all-around security, and security forces are deployed only when they can be positioned so that if detected by an approaching enemy, they will not compromise the ambush. Attack is by fire only, and the enemy is allowed to approach within close range before fire is initiated.

The advantages of the closed triangle formation are ease of control, all-around security, and the enemy can be brought under the fire of at least two automatic weapons, regardless of the direction they approach.

Disadvantages are that it requires a force of platoon size or larger to reduce the danger of being overrun by an unexpectedly large force and that one or more sides of the triangle may come under enfilade fire. The lack of dispersion, particularly at the points, increases the danger from enemy mortar fire.

Open Triangle Harassing Formation

This variation of the triangle ambush enables a small force to harass, slow, and inflict heavy casualties upon a large force without itself being decisively engaged. The assault element is deployed in three teams, positioned so that each team becomes a corner of a triangle containing the killing zone. (See fig. D-12.)

When the enemy enters the killing zone, the team to the enemy's front opens fire on the leading element. When the enemy counterattacks, the group withdraws and the team on the enemy's flank opens fire. When this team is attacked, the team to the opposite flank opens fire. This process is repeated until the enemy is pulled apart. Each team reoccupies its position, if possible, and continues to inflict the maximum damage possible without becoming decisively engaged.

Open Triangle Destruction Formation

The assault element is again deployed in three teams, positioned so that each team is a point of the triangle, 200 to 300 meters apart. The killing zone is the area within the triangle. The enemy is allowed to enter the killing zone; the nearest team attacks by fire. As the enemy attempts to maneuver or withdraw, the other teams open fire. One or more teams, as directed, assault or maneuver to envelop or destroy the enemy. (See fig. D-13 on page D-6.)

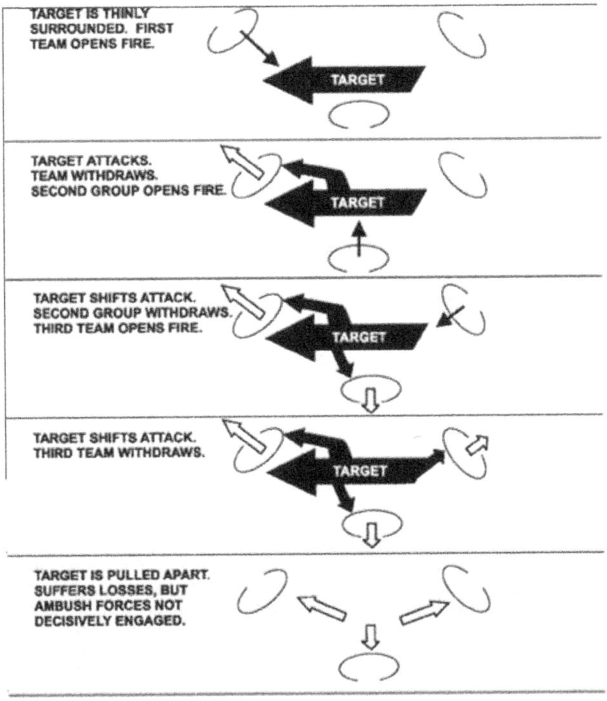

Figure D-12. Open Triangle Formation.

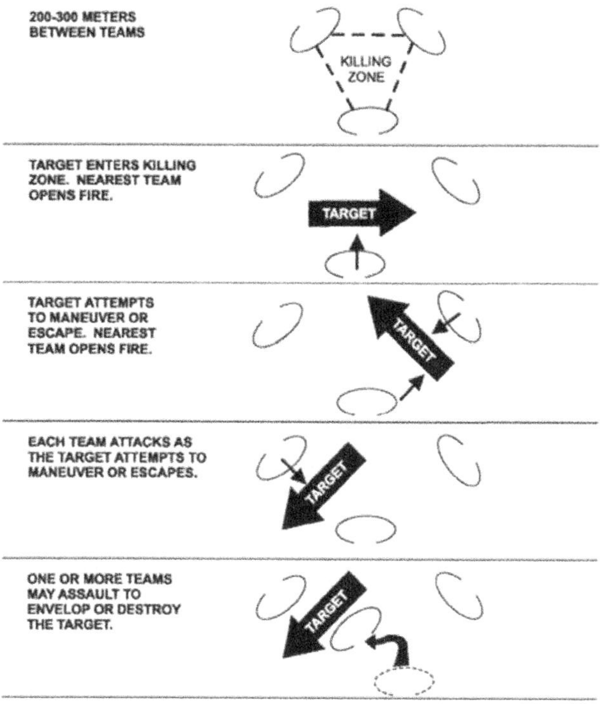

Figure D-13. Open Triangle Formation (Destruction Ambush).

This formation is suitable for platoon-size or larger ambush forces. A smaller force would be in too great a danger of being overrun. Another disadvantage is that control, in assaulting or maneuvering, is very difficult. Very close coordination and control are necessary to ensure that assaulting or maneuvering teams are not fired on by another team. The ambush site must be a fairly level open area that provides concealment around its border for the ambush force.

The Box Formation

The "box" formation is similar in purpose to the open triangle ambush. The assault element is deployed in four teams, positioned so that each team becomes a corner of a square or rectangle containing the killing zone. It can be used for a harassing or destruction ambush in the same manner as the variations of the open triangle formation. (See figs D-14 on page D-7 and D-15 on page D-8.)

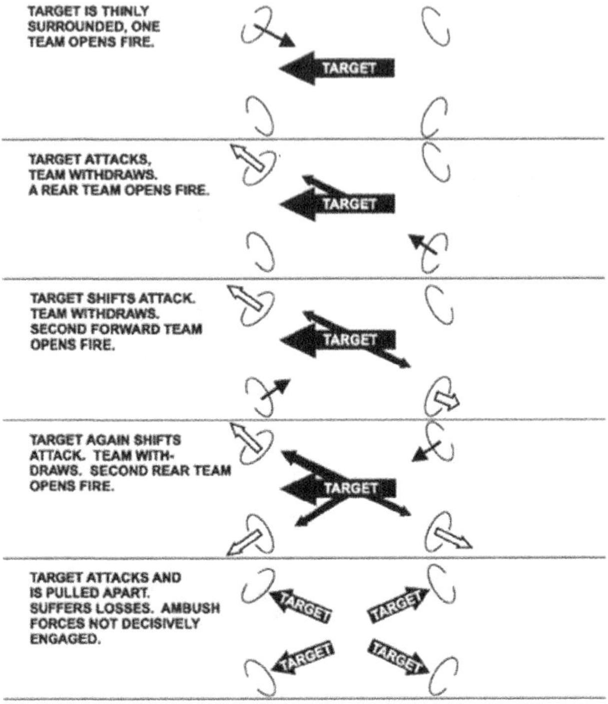

Figure D-14. Box Formation (Harrassing Ambush).

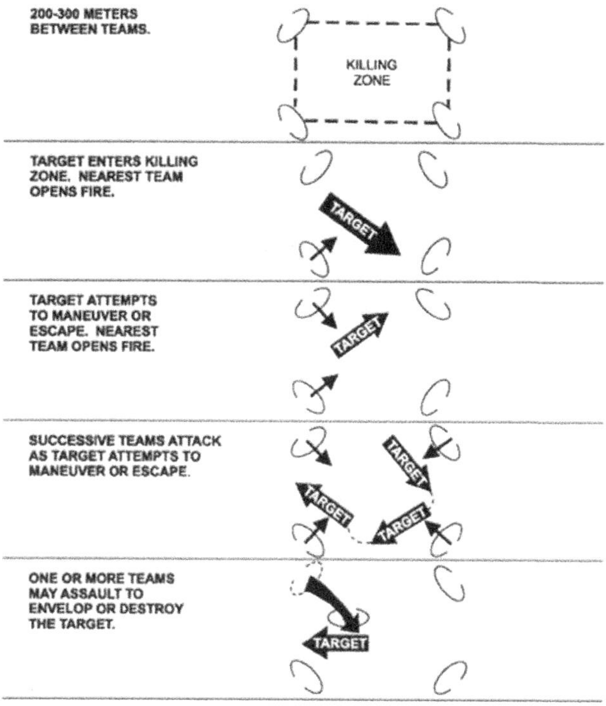

Figure D-15. Box Formation (Destruction Ambush).

Appendix E. Acronyms

A&S . . . . assault and security

CIT . . . . counterintelligence team

EOD . . . . explosive ordnance disposal

ITT . . . . interrogator-translator team

MAGTF . . . . Marine air-ground task force
MEDEVAC . . . . medical evacuation
METT-T . . . . mission, enemy, terrain and weather, troops and support available

NBC . . . . nuclear, biological, chemical
NCO . . . . noncommissioned officer

OTL . . . . overserver-target line

RCA . . . . riot control agents
ROE . . . . rules of engagement

SMAW . . . . shoulder-launched multipurpose assault weapon
SMEAC . . . . situation, mission, execution, administration and logistics, and command and signal
SOP . . . . standing operating procedures
SPOTREP . . . . spot report

TTP . . . . tactics, techniques, and procedures

Appendix F. References

Joint Publication (JP)

1-02 DOD Dictionary of Military and Associated Terms

Fleet Marine Force Manual (FMFM)

6-5 Marine Rifle Squad (proposed MCWP 3-11.2)

Marine Corps Warfighting Publications (MCWP)

3-15.3 Scout Sniping (under development)

3-35.3 Military Operations on Urbanized Terrain (MOUT)

3-41.2 Raids (under development)

Marine Corps Reference Publications (MCRP)

2-15.3A Reconnaissance Patrol Leader's Planning Handbook (under development)

2-15.3B Reconnaissance Reports Guide

3-02C Water Survival Handbook (under development, currently exists as Fleet Marine Force Manual 0-13)

5-12A Operational Terms and Graphics

5-12C Marine Corps Supplement to the DOD Dictionary of Military and Associated Terms

TRAINING CIRCULAR
No. 21-3

HEADQUARTERS
DEPARTMENT OF THE ARMY
Washington, DC, 17 March 1986

SOLDIER'S HANDBOOK FOR INDIVIDUAL OPERATIONS AND SURVIVAL IN COLD-WEATHER AREAS

TABLE OF CONTENTS

*** This publication supersedes TC 21-3, 30 September 1974.**

CHAPTER 4. SMALL ARMS AND AMMUNITION

CHAPTER 5. RATIONS AND DIET

CHAPTER 6. HYGIENE AND FIRST AID

CHAPTER 7. BIVOUAC ROUTINE

CHAPTER 8. INDIVIDUAL MOVEMENT

PREFACE

This handbook is an aid for soldiers in cold climates. However, expertise in dealing with cold-weather operations cannot be gained just from reading this handbook. To develop successful cold-weather field techniques, application training and operational experience are necessary. Additional information is available in FM 31-70, FM 31-71, FM 90-11 (TBP), and FM 9-207.

Military history of operations in northern areas shows that man's success or failure is measured by his regard for the dominant characteristics of a cold-weather environment. For a soldier to become a cold-weather expert, he must learn the techniques and rules needed to fight and live in a cold-weather climate.

The proponent of this publication is HQ TRADOC. Submit changes for improving this publication on DA Form 2028 (Recommended Changes to Publications and Blank Forms) and forward it to Commandant, US Army Infantry School, ATTN: ATSH-B, Fort Benning, Georgia 31905-5593.

Unless otherwise stated, whenever the masculine gender is used, both men and women are included.

CHAPTER 1

Introduction to Cold-Weather Operations

"Soldiers in pairs look after each other; be faithful to your mate; be loyal to your leaders."

Anonymous

1-1. COLD CONDITIONS

COLD-WET

Cold-wet conditions occur when temperatures are near freezing and high and low temperatures cause alternate freezing and thawing. These conditions can be accompanied by wet snow and rain, causing the ground to become slushy and muddy. Soldiers need clothing that has a waterproof or water-repellent, wind-resistant outer layer, and an inner layer with insulation that provide protection in moderately cold weather (above 14°F). **Waterproof footwear is essential.**

COLD-DRY

Cold-dry conditions occur when average temperatures are lower than 14°F. The ground is usually frozen and the snow dry. Low temperatures plus wind increase the need for protection of the entire body. For these conditions, soldiers need clothing that will provide insulation for a wind-chill factor of -80°F. For temperatures below 14°F, white vapor barrier (VB) boots effective down to temperatures of -50°F must be worn.

WIND CHILL

Wind chill is a measure of the combined effects of wind and temperature. Temperature alone does not give a true indication of the impact of the cold. In order to effectively gauge the difference between temperature and the impact of the cold, a wind-chill scale must be used. The chart in Appendix A is a simple and practical guide showing the conditions under which cold weather is dangerous and when exposed flesh is likely to freeze. Soldiers must be familiar with the wind-chill chart to recognize these dangers.

1-2. EFFECTS OF COLD WEATHER ON MILITARY OPERATIONS

Cold has a numbing effect on skin, which hinders task performance. This must be considered when accomplishing such routine tasks as vehicle maintenance, and making and striking bivouac areas (FM 9-207).

When conditions become extreme and the temperature drops, the problem of survival becomes critical. Under these circumstances soldiers may tend to withdraw emotionally, not wanting to leave the warmth and protection of their shelter or tent. The remedy is physical activity. Try to remain alert and active. Execution of all military duties and the fast and proper performance of group tasks are vital.

COCOON-LIKE EXISTENCE

Many soldiers, when wearing several layers of clothing and with their heads covered, tend to withdraw within themselves and assume a cocoon-like existence. When so clothed, many soldiers' hearing and field of vision are restricted; they tend not to notice their surroundings. Their thinking and reasoning become sluggish.

INDIVIDUAL AND GROUP HIBERNATION

This process is characterized by individuals seeking the comfort of sleeping bags, and by the group remaining in tents or other shelter. Guard and security measures may be neglected, jeopardizing the security and safety of the unit.

1-3. OVERCOMING THE COLD

The body must be protected. It must be kept clean, dry, and warm, with normal body processes maintained. Rest and nourishment are vital. The right approach to cold-weather living will keep you healthy and get you through challenging times. There are four basic rules to follow:

KEEP IN SHAPE

Cold-weather clothing and equipment are heavy, and add weight to your normal equipment load. This hinders movement in snow and uses too much energy. The importance of being in excellent physical condition cannot be overemphasized. Your unit must have a tough, challenging, demanding physical training program.

DRINK PLENTY OF WATER

Water may be hard to get, so you may drink less than you need or only when you are thirsty. This will not give you the water needed to avoid dehydration. Drink plenty of water (at least 3.5 quarts per day) to avoid dehydration and fatigue. DO NOT eat snow as a water substitute; it will lower your body's core temperature.

EAT TO KEEP FIT

Regular, nutritious hot food is needed for top performance. You must eat, even when you are not hungry.

KEEP A GOOD ATTITUDE

You will meet many new challenges, but none that cannot be overcome. Your attitude toward the cold will reflect those of your leaders.

1-4. POSITIVE LEADERSHIP AND THE RIGHT ATTITUDE

Most soldiers, after 3 to 4 weeks of cold-weather training, will be able to cope with the environment. Some soldiers may not be as skilled as others, but most will succeed.

Set an example for other soldiers. Winter conditions can be frightening, trying to meet challenges you have never met before. Any mistake could make you a casualty. Aggressive leadership actions that overcome the challenges of the environment are essential to mission accomplishment.

There are two enemies to overcome in northern operations: the enemy soldier and the cold. The first step in defeating either enemy is having the right attitude. The cold can defeat you psychologically if you are not aware of the symptoms of a poorly prepared unit. The following are suggestions on how to defeat the cold:

- If you become depressed or moody, or do not want to talk, force yourself to get involved with others. Do not leave out the weaker soldier.
- If you find it hard to remember things you have been taught, review orders and drills.
- Remind soldiers that their jobs are to close with, engage, and gut the enemy, and that weapons and equipment must be kept in fighting order. Operations in a northern environment can easily become a camping trip with serious consequences. You must avoid this common ambush.
- Cold weather may slow down the following of an order, but it is not an excuse for not doing it.
- Cold weather may make tasks harder and take longer to do, but it will not make them impossible. With knowledge, proper equipment, and proper training you can overcome the cold and be successful in combat.

CHAPTER 2

Personal Clothing and Equipment

"The freezing German troops were reduced to removing clothes from enemy corpses, improvising strawboots, and taking other emergency measures."

Russia, 1942

2-1. INDIVIDUAL CLOTHING

Soldiers should understand the design principles of the military cold-weather clothing system. These principles are: insulate, layer, and ventilate. There are four ways to make these principles work for you: keep it Clean, avoid Overheating, wear it Loose, and keep it Dry. By remembering the "buzz word" COLD, you can quickly check your clothing.

INSULATE

Insulation material reduces the amount of heat loss. By regulating the amount of insulation, an individual can regulate the amount of heat lost or retained.

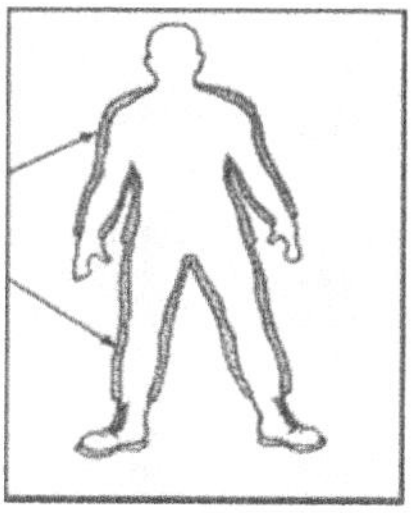

Figure 2-1. Insulation.

LAYER

Several layers of clothing provide more insulation and movement than one heavy garment, even if the heavy garment is as thick as the combined layers. Air pockets trapped between the layers provide insulation and warmth.

VENTILATE

Ventilation helps maintain a comfortable body temperature by allowing excess heat and moisture to escape.

Figure 2-2. C-0-L-D.

KEEP CLOTHING CLEAN

Dirty clothes are cold clothes. Dirt and grease clog the air spaces in clothing and reduce insulation.

AVOID OVERHEATING

Select comfortable clothing and do not overdress. In cold weather, it is better to be a little cool rather than risk being overheated.

WEAR IT LOOSE

All items of the cold-weather uniform are large enough to allow wearing of the appropriate number of layers. Your field jacket may appear too large without all of the layers designed to fit under it. If the uniform items do not fit loosely, the insulation that keeps you warm will be greatly reduced.

KEEP IT DRY

It is important to keep clothing dry, since wet clothing conducts heat away from the body more quickly than dry clothing. Moisture soaks into clothing from two directions: the inside and the outside.

From the outside you must guard against melting snow; from the inside you must guard against sweating. Brush snow and frost from clothing before entering any heated shelter or vehicle, and avoid overheating.

2-2. COLD-WET VERSUS COLD-DRY

As mentioned in Chapter 1, temperatures that are above 14°F (-10° C) are considered cold-wet conditions; temperatures consistently below 14°F will cause cold-dry conditions. Cold-wet conditions usually involve a freeze/thaw cycle, often accompanied by rain or wet snow. This type of weather is hard to dress for, since each soldier must battle the elements to keep his clothing dry. During cold conditions, the ground is usually frozen and the snow is dry. Athough the temperatures during the cold-dry conditions may be lower than during cold-wet conditions, it is easier to stay warm.

COLD-WET		COLD-DRY
	OR	
DOWN TO 14°F		BELOW 14°F

Figure 2-3. Cold-wet versus cold-dry.

2-3. THE COLD-WEATHER UNIFORM

There are many layers of clothing that must be worn in a cold weather environment. The first layer of clothing is the T-shirt and shorts. The second layer is a set of thermal underwear, which consists of an undershirt and underpants. You should wear only the clothing needed to keep comfortable and warm.

The scissors-type suspenders must be the last thing next to the outer layer, or else the layers cannot be tucked inside the trousers. The underpants and all succeeding layers of trousers are supported from the metal hooks on the suspenders. The suspenders allow the underpants and trousers to be worn loosely at the waist so that circulation and ventilation are not restricted.

Wear the wool shirt outside the trousers, which provides another method of ventilation. You can increase the waist size of this shirt by removing the stitching in each dart at the sides of the shirt.

Wear the field trousers as an outer garment in the cold-wet environment. In the cold-dry environment, wear the nylon liners inside the field trousers during periods of inactivity. On the inside of each cargo pocket on the trousers is a tape that runs through a hole on the side of the pocket. The tape ties around the thighs and keeps the cargo in the pocket from rubbing and irritating the legs. For warmth and to prevent entry of snow, tie the drawstrings around your ankles outside the boots into the boot lacing, and tuck the bowknots under the trousers.

The field jacket is worn with the liner in both cold-wet and cold-dry areas. Wear the coat and liner directly over the wool shirt. Remove the liner in mild weather or when activity increases.

In a cold-dry environment the parka is worn with a liner. The waist is adjustable for ventilation and warmth with a drawcord, and the skirt has a "split-tail" that can be tied tightly around each leg for warmth. During a period of severe cold, when activity is less, wear the parka with liner over your field jacket with liner. Remember: do not overdress or you will sweat.

Do not dry nylon on drying lines too close to the heat; too much heat will melt the nylon, destroying its insulating value.

Keep the head covered to stay warm; a warm head even warms the feet. Wear the cold-weather cap or balaclava under your helmet or under the hood to your field jacket. When wearing the cold-weather cap as an outer headpiece, the lower flap may be folded up over the top and fastened with hook-pile tape straps in front. In a cold-dry environment, wear the winter hood over the cap or the helmet. Adjust the hood by using the drawstrings on either side of the hood and the elastic strip at the back. To provide minimum protection for the face from the wind, bend the wire in the fur cuff of the hood so that you have a small opening for your face. This allows visibility but less protection.

The feet are hard to keep warm and dry, so protect them. The insulated boots (black for cold-wet conditions and white for cold-dry conditions) are warm and waterproof. The white insulated boot is the best cold-weather footwear today. When you are active, your feet will sweat and, because of the waterproof rubber lining, the moisture will remain in the socks or in the bottom of the boots. The feet, even if damp, will stay warm because the heat given off by your body will remain inside the boots. If the boots are worn for long periods, the feet become white and wrinkled when you remove the boots. This is no cause for alarm. Drying, massaging with foot powder, warming and putting on dry socks will correct this. If you are inactive for long periods, your feet may become cold. Exercises such as knee bends, stamping your feet, running in place, or wiggling your toes will make them warm. Elevate the feet whenever possible to aid blood circulation. Don't wear socks that are too tight, and wear only one pair of cushion-sole socks at a time. Change your socks at least twice a day. Most important, stay active.

Care of your boots is important. Wash boots inside and out with soapy water at least once per month or as required by unit SOP you want your feet to keep warm, you must keep the insulation dry. Inspect your shoes often and repair any holes or punctures immediately. Temporary repairs can be made by patching the hole with any kind of tape; even chewing gum can be used. Seal the hole quickly to prevent the insulation from becoming wet. If one of your boots feels heavier than the other, the insulation is probably wet and they should be turned in for new ones.

Insulated boots must be tested annually to ensure they are waterproof and have less than 0.01 percent moisture in the insulation. The air valve on the outside of the boot is used to equalize pressure when worn at high altitudes and adjusted again upon return to low altitudes. To equalize the pressure in the boot, open the valve for a few seconds and immediately close it. The valve should be kept closed at all other times to prevent moisture from entering. Never open the valve unless the boot pressure around your foot causes discomfort.

You can wear overshoes over the leather boots on wet or muddy terrain. The overshoes are not as warm and dry as the insulated boots. Make sure your socks are dry and that your boots are loosely laced so as not to restrict circulation.

2-4. EXTENDED COLD-WEATHER CLOTHING SYSTEM (ECWCS) (TEST)

The ECWCS is issued as a one-for-one replacement for the current standard cold-wet clothing system. System components use the most recent materials and technology. The result is a reduction in weight and bulk from the current cold-wet clothing system without loss of protection from the cold. The ECWCS uses the layer approach that draws moisture away from the body and should be worn as discussed below.

FIRST LAYER

This consists of polypropylene undershirt and shorts. They are worn next to the skin. They are designed to draw moisture away from the skin, thereby reducing conductive heat loss. You must make sure that no other item is placed between these items and your skin.

SECOND LAYER

The second layer contains a synthetic fiber pile (SFP) shirt and polyester batting trouser liners. The SFP shirt and trouser liners provide insulation. The shirt is designed to dry quickly if it becomes wet and to provide a high degree of insulation. It should not be worn as an exterior garment in wet or windy conditions, since the loose weave provides little protection against wind chill. If the shirt gets wet, it will reduce the insulation value.

The trouser liners are designed to be worn during little or no activity. They should not be worn during heavy activity such as skiing or snowshoeing.

THIRD LAYER

This consists of polyester batting field jacket liner and field trousers. The jacket liner should only be used for extra insulation during periods of little activity. This liner is designed to be used as a separate jacket; it should not be attached to the parka. The trousers provide a durable wind-resistant, water-repellent layer to protect the insulating underpants and trouser liners.

FOURTH LAYER

Windproof, waterproof parka and trousers make up this layer. The parka and trousers provide windproof and waterproof shell garments to protect the first three layers.

The parka should be the outer garment, which is always worn when doing any activity outside a shelter. It is waterproof, yet the fabric is breathable so that body moisture can escape. The hood is designed to give full protection from wind and snow and does not prevent your seeing the area around you. It has underarm zippers to allow air to circulate during heavy activity. The trousers are made of the same material as the parka and are very warm. They should be worn only when necessary to prevent the field trousers from becoming wet.

FIFTH LAYER

This consists of standard camouflage overwhites.

2-5. TIPS ON WEAR AND MAINTENANCE OF CLOTHING

The previous discussion on the layers of cold-wet and cold-dry uniforms describes how each system functions. In some units, the cold-dry field trouser liners are often substituted for the wool field trousers of the cold-wet system and function well in a cold-wet environment. Your commander and unit standing operating procedure (SOP) will determine what uniform and components will be worn and carried in the field.

DRYING WET CLOTHING

Hang items separately from drying lines inside the tent. Do not place anything too close to a stove or over steaming pots; steam will cause further dampness.

DRYING DAMP CLOTHING

Damp items can be hung on the rucksack during movement. Pin damp socks to load-carrying equipment or shirt during movement or just inside the last layer of clothing. Do not place damp clothing in sleeping bags. This will only make sleeping bags wet, and they are harder to dry than clothing.

Small items can be placed between the sleeping bag and the sleeping mat under your legs. When washing clothes without soap, dry-rub and air clothing and sleeping bags.

Never warm your feet next to an open fire. The boots will melt and burn because of their insulation before the feet become warm.

Try never to wear VB boots in a sleeping bag. Sometimes the tactical situation is such that boots must be worn while sleeping. If so, the feet must be dried and massaged with foot powder, and dry socks must be put on. Wearing the VB boot while sleeping does not let your feet dry out. After VB boots have dried, place them underneath your sleeping bag on top of your sleep pad when temperatures are below 0°F.

2-6. COLD-WEATHER EQUIPMENT

The sleeping equipment consists of a sleeping bag, an insulated sleeping pad, and a waterproof bag. There are two types of sleeping bags:

(1) Sleeping bag type I, intermediate cold, temperature range +45°F to +10°F, consists of a mummy-shaped bag with hood, insulated sleeping pad, and a waterproof bag.

(2) Sleeping bag type II, extreme cold, temperature range of +10°F to -50°F, is also available.

Whatever clothing is necessary must be worn with both bags to stay warm and comfortable at the lower temperature ranges. As a minimum, long underwear, sleep shirt, dry socks, and head protection should be worn while sleeping.

The insulated sleeping pad has replaced the pneumatic mattress. Use the insulated sleeping pad under the sleeping bag for warmth and comfort and to keep your sleeping bag dry.

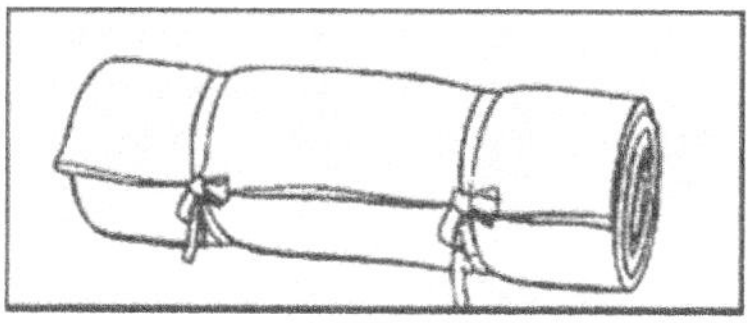

Figure 2-4. Sleeping pad.

The bag must be kept dry if it is to keep you warm. Sleeping bags become wet from outside moisture, from sweating on the inside of the bag, and from water escaping from the mouth or nose into the sleeping bag.

To keep outside moisture from wetting the bag, place protective insulation, pine or spruce boughs, a poncho, or an insulated sleeping pad under the bag. Avoid sweating by wearing the least amount of clothing to keep warm inside the bag, and by using the proper sleeping bag for the temperature conditions. Do not breathe into the bag or moisture will collect and wet the bag or form ice crystals. If your face gets cold, cover it with a towel, muffler, balaclava, or sleeping hood. The sleeping shirt with hood is specifically designed for use with the types I and II sleeping bags.

When it is not in use or being carried, open the sleeping bag wide so that fresh air can get into it. Turn it inside out so that the sun and wind can dry it. Always shake the bag out and fluff it up before and after use to provide for more trapped air between the layers of the wind-resistant/water-resistant outer shell.

Always carry your sleeping bag in the waterproof carrying bag. This protects the bag and helps keep it dry. Ensure that the top bag is securely tied to prevent snow or other moisture from getting inside and wetting the sleeping bag. The following suggestions will help you keep warm and comfortable in the sleeping bag:

- Eat just before getting into the bag. This gives you energy and helps you to stay warm.
- Always relieve yourself just before going to bed. The act of getting out of a warm sleeping bag to go outside can be a chilling experience.

PACKING YOUR RUCKSACK

When you load your rucksack for skiing or climbing, pack heavy items at the bottom and next to the frame. This places the weight on your hips, which is necessary for good balance. Place hard or sharp objects inside the load where they will not rub on the bag and your back. When walking or snowshoeing, raise the center of gravity and take more of the load on your shoulders. Articles that are often needed are placed in the outside pockets for easy reach. Keep maps and other flat objects in the flap pocket.

ADJUSTMENT

Adjust the straps so that you can move the top part of your body and swing your arms freely. Adjust the shoulder straps so that the lower back straps fit just above the belt line.

2-7. LOAD-CARRYING EQUIPMENT

The all-purpose lightweight individual carrying equipment (ALICE) rucksack comes in sizes medium and large. Make sure you are equipped with the large ALICE in a cold environment.

In cold-weather operations, you should follow the packing list set forth in your unit SOP to organize your rucksack. The main consideration is to avoid having to unpack during halts to get frequently needed items. The following illustrates a method of organizing your rucksack:

- **Small external pockets.** These contain small high energy foods (candy, cereal bars) to be eaten while on the move.
- **Large external pockets.** These contain rations to be eaten for morning and evening meals. Take extra socks, scarves, and a spare cap. The openings behind the pockets can be used for skis or other items.
- **External attachment points.** The sleeping pad is attached to the bottom of the pack or under the top flap (wrapped inside sandbags) and other combat items as required.
- **Top flap.** It contains camouflage overwhites and pack cover.
- **Main compartment.** Place the sleeping bag in the bottom of the rucksack. Spare clothes are placed in the upper half of the compartment for easy reach.
- **Camouflage.** A white cotton cover is provided that covers the entire rucksack, excluding the shoulder straps. Use it when wearing your overwhites.
- **Care and maintenance.** Check your rucksack for damages before and after an operation. Repair damages immediately or turn in through supply channels for a replacement.
- **Emergency quick-release of the rucksack.** When quick removal of the rucksack is needed, unsnap the fastener. Pull down on the quick-release fastener on the left shoulder strap, shift the rucksack toward the right shoulder, and let it drop to the ground.

- **Cargo support shelf.** The rucksack can be converted to a cargo carrier by removing the combat pack and replacing it with the cargo support shelf. You can carry square, rectangular, or box-type items such as radios, water cans, and fuel cans.

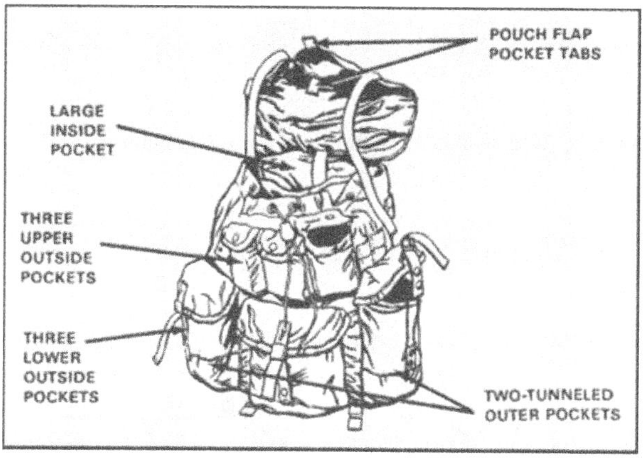

Figure 2-5. Load-carrying equipment.

2-8. OVER-THE-SNOW MOVEMENT EQUIPMENT

Before you are instructed to use snowshoes or skis, your chain command will first determine if movement by foot is possible. In shallow snow (8 inches or less) or heavily crusted snow, movement on foot for inexperienced skiers provides better maneuverability with the least effort.

SNOWSHOES

Snowshoes consist of a white magnesium frame laced with nylon-coated steel cable. They are about 48 inches by 12 inches weigh 3.3 pounds with bindings. When properly adjusted, the toe of the boot will be about halfway across the opening in the snowshoe; the toe strap should be firm, but not tight enough to restrict circulation; and the heel straps should permit some side-to-side movement. When used correctly, snowshoes provide better mobility than walking. In most instances, skis are preferred over snowshoes. When using snowshoes, you cannot back up or swim; be careful when moving over thin ice. The traction needed to ascend or descend with snowshoes can be increased by waxing. The use of ski poles with snowshoes aids movement and reduces fatigue.

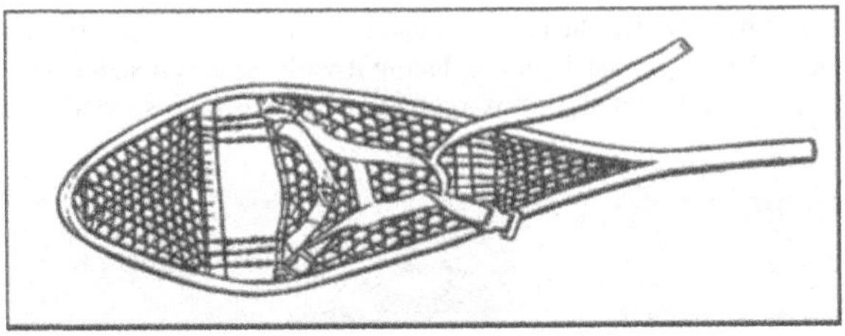

Figure 2-6. Snowshoe and snowshoe binding.

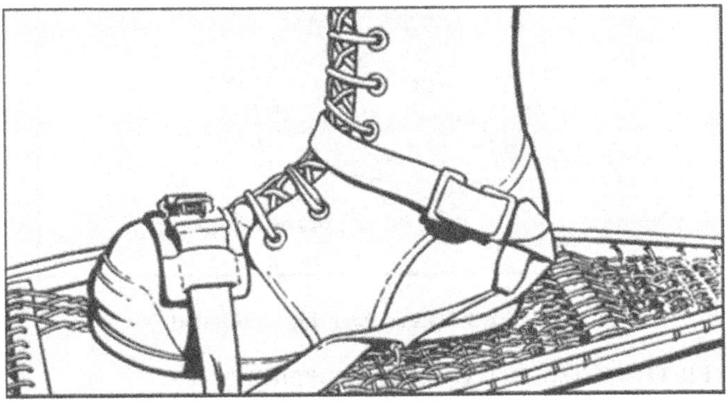

Figure 2-7. Snowshoe binding attached to VB boot.

SKIS

The standard military skis are about 208 centimeters (82 inches) in length, and have a hole in the tip for towing, a groove in the heel for attaching mohair climbers, and metal edges.

SKI POLES

Ski poles can be very useful in steep terrain. The military ski pole consists of a steel shaft with rubber handgrips and baskets, and a leather wrist strap. The poles come in three lengths: 130 centimeters (51 inches), 137 centimeters (54 inches), and 147 centimeters (58 inches). When fitting the poles, you should stand on the floor and place the pole at your side; it should come to the point of your shoulder (about 1 inch below the top of the shoulder).

MOHAIR SKI CLIMBERS

The mohair climbers are attached to the bottom of the skis for added grip when climbing steep slopes. The climbers consist of a cotton-backed mohair pile and are attached to the skis by five web straps.

2-9. MISCELLANEOUS EQUIPMENT

SUNGLASSES

Polarized sunglasses should always be worn on bright days when the ground is covered by snow. Failure to wear the sunglasses, even on overcast days when the need seems less, may result in blindness.

PROTECTIVE MASK

There is an M4 cold-weather winterization kit available for your protective mask. Make sure you install this winterization kit for operations in the northern environment.

CANTEENS

In cold weather the plastic canteen will freeze very quickly if carried in the standard fabric carrier. Since the canteen cannot be placed directly on heat, there is no quick method for thawing. Whenever possible, the canteen should be carried in one of the interior uniform pockets, or wrapped in clothing and placed in the rucksack to ride against your back. In extreme cold, do not fill the canteen over two-thirds full; this allows for expansion if the water freezes.

The insulated canteen is a one-quart capacity stainless steel vacuum bottle. It is provided with a cotton duck cover and a stainless steel cup. A new arctic canteen will keep water from freezing at -40°F for 6 hours. The following are some tips in using the arctic canteen:

- Fill the canteen using your canteen cup.
- Fill to within half an inch of the neck.
- Do not drop or dent.
- Put it in your sleeping bag at night.
- Carry the canteen in an M60 bandoleer near your body.

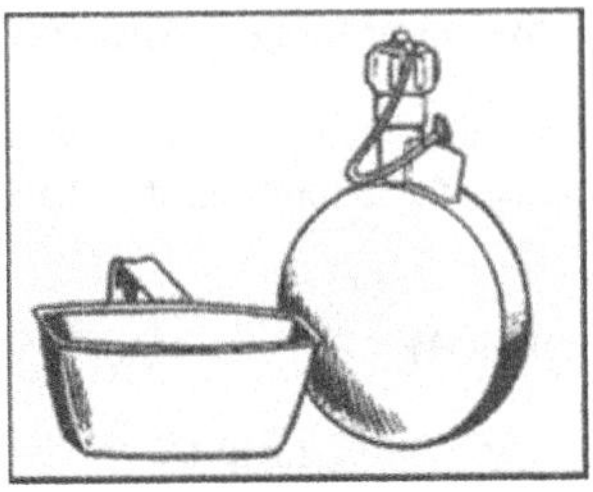

Figure 2-8. Insulated canteen with cup.

CHAPTER 3

Tents and Heating Equipment

"A well-sheltered defender enjoyed a marked advantage over his exposed attacker..."

Ironside, 1918

3-1. GENERAL

To conduct successful military operations in cold weather and maintain a high level of combat efficiency, you must have heated shelters. Tents and stoves are a vital part of cold-weather equipment.

3-2. TENT GROUP EQUIPMENT

The following is a list of the tent or ahkio group for an infantry squad:

1 Ten-man tent with liner
1 Yukon stove
1 Metal or plywood stove base
2 Five-gallon gasoline cans (one with white gasoline per platoon)
1 Five-gallon water can (filled)
2 Cases, meals, ready-to-eat
1 Climbing rope (120 feet)
1 Box of candles
2 Squad cook sets
2 One-burner squad stoves
2 Axes
2 Saws
1 Hatchet with sheath
1 D-handle shovel
1 Sledgehammer
1 Pick
1 Tripod for gas can (set)
3 Pairs of snowshoe bindings (spare)
1 Lantern (extra mantles)
1 Box of soda bicarbonate (fire extinguisher)
1 Roll of soft wire
5 Plastic bags
3 Rolls of toilet paper

3-3. AHKIO

The ahkio is the squad's method of transporting the tent group equipment in the winter. It is a 200-pound capacity fiberglass sled that can be used for transporting tent group equipment, for carrying weapons and ammunition, as a firing platform, and for the evacuation of casualties.

3-4. PACKING

When packing the ahkio, the load must be evenly distributed. Place heavy equipment on the bottom and slightly to the rear of center. Loading lighter equipment toward the top will prevent the ahkio from becoming top heavy. Tools such as shovels, axes, and saws are packed at the sides for easy access in trailbreaking; however, caution must be taken to ensure that these items do not stick out. The canvas cover is folded over the load. To keep the load from shifting and to prevent snow from getting under the cover, lash the load tightly by crisscrossing the lashing rope from the lashing rings on one side of the sled to the rings on the other side. An additional ahkio, issued on a basis of one per two 10-man tents, should be used to carry camouflage nets and support systems. In very wet or dry snow, the movement of an ahkio is easier when wax is applied to the running surface.

3-5. TEN-MAN ARCTIC TENT

The 10-man arctic tent with liner is a six-sided pryamidal tent supported by a telescopic pole. The tent is designed to accommodate 10 men and their clothing and equipment. It will hold additional personnel by leaving packs and equipment outside the tent, or by lowering the center pole, which allows the sidewalls to spread out to cover more ground surface. The tent has two doors, which permits joining of tents for access from one tent to another. A snow cloth is attached to the bottom of the sidewalls for sealing the tent to the ground. The snow cloth should always be to the outside of the tent with pine or spruce boughs underneath, or it may freeze to the ground. This allows easy escape in case of an emergency if the doors are blocked. The tent is ventilated by four built-in ventilators on opposite sides and near the top of the tent.

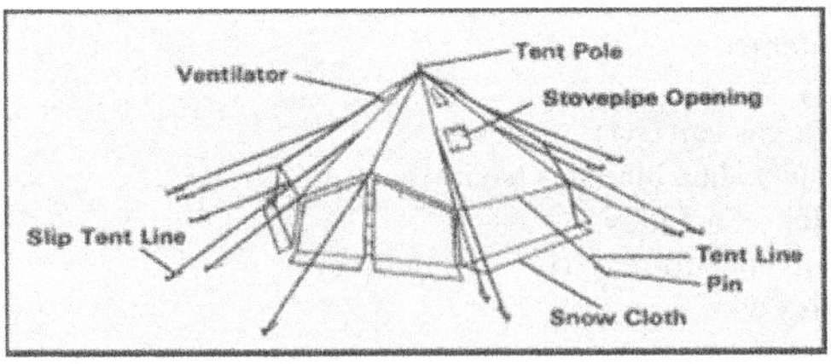

Figure 3-1. Ten-man arctic tent.

3-6. BUILDING ARCTIC TENTS

The shelter is easily built by four trained soldiers. Consider the following when pitching or striking tents in snow or on frozen ground.

Snow should be cleared to the ground surface to obtain a lower silhouette and gain advantage of ground temperatures, which are generally warmer than upper air temperatures. Enough pine or spruce boughs should be placed on the tent and floor, and under the snow covers, for insulation and comfort. If an insulating material is not used, the heat of the stove will thaw the ground insulation in a sea of mud. The snow covers will freeze to the ground, preventing escape from the tent in case of an emergency. When it is impractical to remove the snow to the ground level, an adequate tent site can be prepared by packing the snow with skis or snowshoes until a firm base is provided for pitching. The center pole is placed on a log or other suitable support to keep it from sinking into the snow.

In open terrain, with a strong wind, it will be necessary to build a 1-meter-high snow wall on the windward side of the tent to protect it from the wind. The tent is pitched with the entrance angled about 45 degrees from the downwind side. When the tent is set up, the snow cloth around the bottom of the tent should be spread out so that snow or logs can be placed on it to help in anchoring the tent.

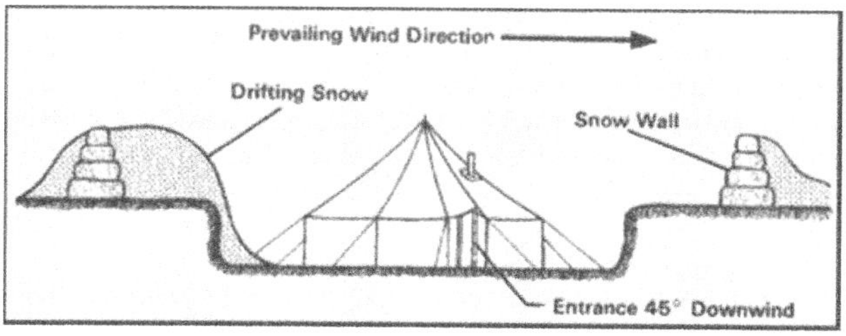

Figure 3-2. Windbreak.

Driving tent pins into frozen ground is impossible. Tents may be pitched and anchored by attaching the tent lines to trees, branches, logs, or stumps. If these are not available, holes are dug into the snow for the purpose of using "deadmen." This is done by digging a hole into the snow large enough to insert a pole or log about 1 meter long with the tent line attached. The hole is filled with snow and well packed. The packed snow freezes, securely anchoring the tent. In powder or dry snow, water would be poured into the hole.

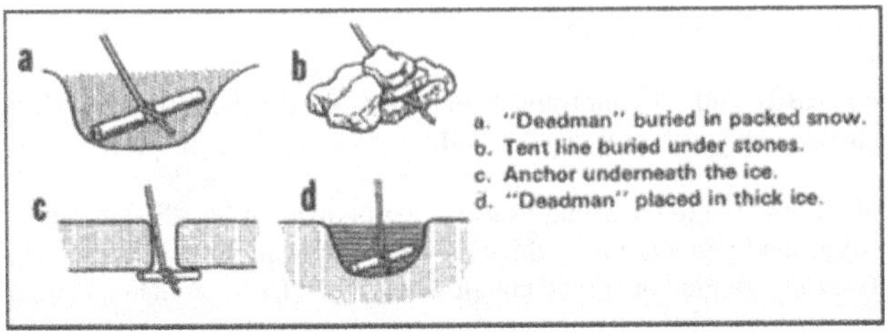

Figure 3-3. Deadmen.

The smaller tent (five-man) provides the commander more flexibility in distribution of his forces. You can never be assigned a tactical mission in the arctic without provisions for a shelter. This tent enables a smaller unit to be assigned an independent mission, such as patrol or reconnaissance mission. It is also six-sided, pyramidal, and supported by a telescopic tent pole. It is designed to accommodate four to five men and their clothing and equipment. The tent can provide shelter for a rifle squad or other similar unit when rucksacks are placed outside the tent.

3-7. INSTRUCTIONS FOR PITCHING TENTS

Instructions for pitching tents are sewn securely to the tent just inside the door. Although the instructions are easily read and followed, there is one thing that should be remembered when pitching either tent. After the tent is spread out flat on the ground, the six corner eave lines should be staked out and loosely secured to the stakes before the tent is raised. These lines are yellow for easy identification. The stakes or tie-down points for these lines should be located about 2 meters from the tent. After this is done, proceed with the tent pitching as indicated in the instructions. The door eave line should be propped up by placing the line over a pole (tree branch or other object higher than the door entrance) and then stake the line out to a pike or tie-down point. This procedure keeps the door from sagging and makes the zipper work better. Drying lines should be attached to the liner inside the tent to allow drying of clothes.

In winter the tent will normally be covered with ice and snow, which must be removed before the tent is struck. This can be done by shaking the tent or by beating it with a mitten or stick. If the snow cloth is frozen to the ground, the snow and ice around it must be removed by chopping or shoveling. This must be done carefully to avoid damage to the snow cloth. One way to do this is to ease the shovel between the cloth and the ground and gently pry the cloth away from the ice.

When the bottom of the tent is properly sealed and the door is zipped, moisture will form on the inside of the tent and collect on clothing and equipment. Carbon monoxide and fumes from the stove can accumulate to a dangerous degree. To avoid this, ensure that the ventilators near the peak of each tent are open.

The roof of the shelter must be steep enough to shed snowfall. With both the 5-man and 10-man tent, the area that will become the floor of the tent is dug, packed, and insulated with pine, spruce boughs, or similar material. Snow walls should then be built to protect against wind.

3-8. YUKON STOVE

The Yukon stove is used to heat the 10-man, 5-man, and GP small tent. In addition to providing heat, the top surface of the stove and the area beneath the stove may be used to heat rations and water. The stoves use standard MOGAS but may also be operated with diesel fuel or JP-4 without modification. NEVER MIX FUELS. When solid fuels (wood, coal, etc.) are used, the stove must be modified by removing the burner unit, closing the burner opening, and turning over the wire grate so there is space below for draft and ashes.

WARNING

DO NOT USE A SPARK ARRESTER ON THE M1950 YUKON STOVE.

A flue cap, issued as a component part of the Yukon stove, shields the stovepipe top from the wind and prevents the wind from blowing smoke or gases down the pipe into the stove and tent. The flue cap is equipped with three 4 1/2-meter lines that are used to steady the stovepipe and anchor it in high winds. These guy lines must be anchored to the tent or tent ropes, not to the ground or nearby trees. The stovepipe hole cover must be rolled up and tied off to prevent nylon ties from touching the stovepipe.

A firm base for the stove must be provided to keep it from melting down into the snow. The base should be 35 centimeters by 70 centimeters and be made of lightweight metal or salvaged plywood. If plywood is used, it should be covered with metal or aluminum foil. If those materials are not available, a base can be made out of small logs about 10 centimeters in diameter.

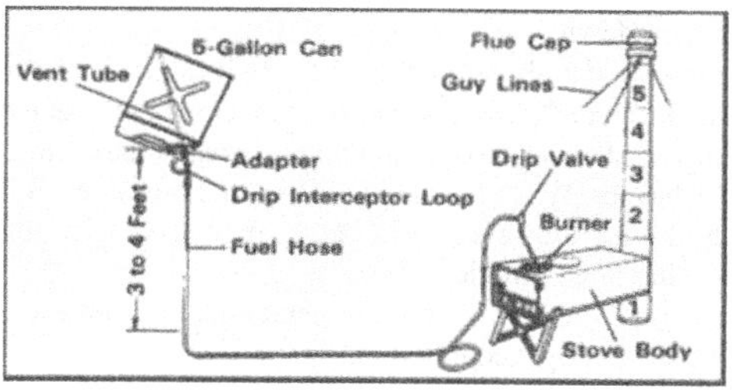

Figure 3-4. Yukon stove.

A simple stand for holding the fuel can is built by obtaining three poles 2 meters long. The poles are tied about two thirds of the way up and then spread out to form a tripod. The fuel can should be at least 1 meter higher than the stove. If the fuel can is wobbly or if there is a wind, the can should be tied to this tripod. Make sure the can is tilted so that the air is trapped in the uppermost corner.

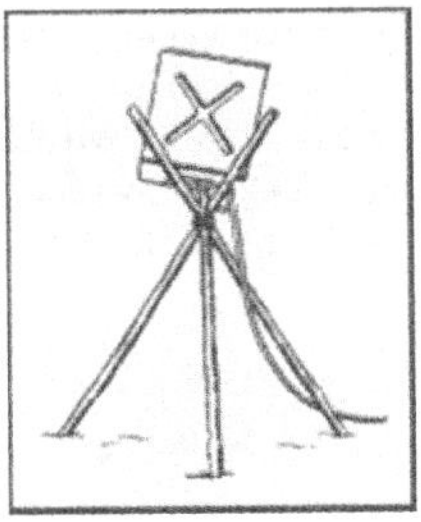

Figure 3-5. Tripod for gas can.

3-9. PRECAUTIONS

The following precautions must be observed when the Yukon stove is being used:

- All stovepipe connections must be tight, and the tent stovepipe shield properly adjusted. The stove must be level to ensure that the burner assembly will spread the flame evenly throughout the stove.
- Ensure that the fuel line is made of natural rubber and suitable for extreme-cold operation. It must be protected so it cannot be accidentally pulled loose and must have a drip intercept loop to prevent fuel from running down the line and into the tent.
- Rate of fuel flow must be checked regularly. It will change with the fuel level and will need adjustment. All fuel supplies must be kept outside the tent.
- The stove should never be left unattended when it is burning. Allowing the stove to become too hot may cause the stove body to become overheated and warped and burn the tent down.
- If the flame is accidentally put out, or if the fuel can is being changed, the drip valve must be turned off. When the stove has cooled, excess fuel inside the stove must be removed. Two or three minutes must pass so gas fumes escape before relighting the burner. When burning gasoline, the burner must be cool before relighting the stove. If the stove is lit before the burner has cooled, the fuel will vaporize, causing it to explode.
- If the tactical situation does not require personnel to be awake 24 hours a day, turn the stove off during sleeping time.

The following are precautions when **burning wood or coal**:

- Fuel should be fed in small amounts until there is a hot bed of coals.
- The stove should not become overheated.
- Gasoline should not be poured on the wood or coal inside the stove to hurry starting the fire.
- Ashes should be removed regularly and not allowed to accumulate higher than the bottom of the grate.
- NEVER MIX FUELS.

3-10. SQUAD STOVE, M1950

The one-burner squad stove is a cooking and heating unit for two to five men. It is used when operating in an isolated or forward area when separated from the ahkio group. The stove is small, compact, and lightweight, and it will operate on either MOGAS or white gasoline. The initial pressure for its operation is generated by a few strokes on the hand pump. Thereafter, the pressure is generated by heat from the burner evaporating the fuel in the stove tank.

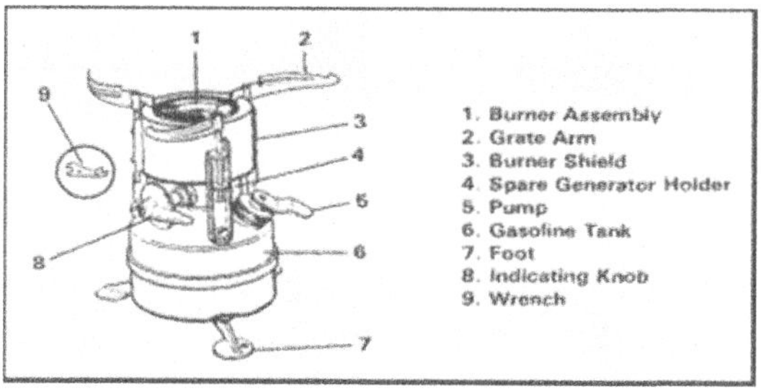

Figure 3-6. Gasoline cooking stove, M1950.

To operate the squad stove, do the following:

- Snap out the three feet and set stove on a level surface.
- Raise the grate arms and turn them outward so that the inner tabs can be engaged in the slots in the burner shield.
- Close the valve by turning the indicating knob clockwise to the OFF position.
- Unscrew the pump caps and remove the pump assembly.
- Fill the gasoline tank three-fourths full.
- Wipe off any gasoline that may have been spilled during filling and replace the pump cap.
- Pump 10 strokes to build up initial pressure.
- Turn the indicating knob to the ON position for 3 seconds. This allows the gasoline to rise through the generator to the burner and preheated cup.

- Turn the indicating knob to the OFF position to prevent excess fuel from accumulating in the burner assembly.
- Light the wick and allow the flame to burn for about 3 minutes or until the flame turns blue.
- Turn the indicating knob to the LIGHT position. Have a second match ready to relight the burner immediately if the flame goes out.
- When the flame in the burner turns blue, turn the indicating knob to the ON position.
- Pump an additional 10 strokes to maintain pressure within the tank. The stove is now ready for use.
- If the flame burns unevenly or dies during operation, turn the valve to the OFF and ON position several times, relight quickly, and pump several additional strokes.
- To turn the stove off, turn the indicating knob to the OFF position and allow the flame to go out.

CHAPTER 4

Small Arms and Ammunition

"Rifles and machine guns and even the breechblocks of artillery became absolutely rigid ... and steel parts cracked."

General Rendalic, Eastern Front, 1942

4-1. GENERAL

Cold weather will greatly affect the operation of your weapon. You must be aware of these effects to ensure employment of your weapon under poor weather conditions.

4-2. CARE, CLEANING, AND MAINTENANCE

Your weapon will work properly under extreme cold conditions if you take care of it. Normal temperate zone lubricants thicken in cold weather, and stoppages or sluggish actions of your weapon will result. During the winter, disassemble your weapon completely, clean it thoroughly, and remove all lubricants. Oil the entire weapon with a light coating of cleaner lubricant preservative (CLP).

CAUTION

DO NOT USE CLP ON A MORTAR OR ANY COMPONENT PART OF A MORTAR.

Figure 4-1. Weapons will 'sweat" when taken in to a heated shelter.

A main problem is to keep snow and ice, which may disable your weapon, out of the working parts, sights, and barrel. When the weapon is not in use, keep it out of the snow either in an improvised weapon stand or in a tripod arrangement. If this is not possible, stand it up with the butt end in the snow.

Condensation ("sweating") will form on the weapon when you take it from the cold into a heated shelter. The condensation will continue for about 1 hour, after which you should clean the weapon. If you don't, the condensation will freeze when you take the weapon outside, and it may not fire. The weapon must be kept as near the floor as possible to lessen condensation. If parts of the weapon do become frozen, warm them and move the working parts slowly until they are unfrozen. If you can't warm your weapon, remove all ice and snow and move the parts gently until action is restored. The use of CLP will prevent malfunctions by eliminating condensation.

When firing, do not let the hot parts of your weapon come in contact with the snow. The snow will melt and, as the weapon cools, will form ice. When changing barrels, don't lay them on the snow.

To prevent bipods and tripods from sinking into the snow, improvise for added flotation. This is done by placing them on snowshoes, ski poles, branches, ahkios, etc. Baseplates should be on solid ground or sandbags to prevent sinking into the snow and damage to the weapon.

Figure 4-2. Bipods and tripods will sink in the snow.

AMMUNITION

Small arms ammunition should be kept at the same temperature as your weapon. Ammunition clips and magazines must be cleaned of all oil and preservative and checked often. All ice, snow, and condensation must be removed. Ammunition boxes must be kept closed in order to keep out snow and ice.

Stored ammunition should be kept in packed boxes, raised off the surface by placing them on boughs, cardboard, or other similar material, and covered to protect them from the snow. Ammunition should be well marked for easy location and identification in case it becomes covered with snow.

Ammunition resupply may become restricted. Everyone must be aware of the necessity for ammunition economy and fire discipline. Loaded clips, magazines, or single rounds dropped in the snow become quickly lost; therefore, careful handling of ammunition is essential. Never waste your ammunition. Make every round count.

When contact with the enemy is imminent, troop compartments and aircraft should be kept as close to freezing as possible to avoid condensation. However, many munitions have limitations at lower temperatures, which must be recognized in training.

CHAPTER 5

Rations and Diet

"I ate my boots and a pair of pants ... tea and seal-skin gloves for dinner."
Greely, Arctic Expedition, 1883

5-1. GENERAL

Most of what we eat and drink is used in maintaining our body heat, while only a small portion is used in producing energy for physical work. You must ensure adequate caloric intake in cold-weather operations. About 4,000 calories per day are necessary for personnel performing physically demanding work in the cold. Efficiency may drop rapidly if this level is not maintained.

The body loses liquid at a very fast rate in arctic conditions, regardless of how carefully you adjust and ventilate your clothing. The exertion of movement on foot, preparation of bivouac sites and defenses in the snow, etc., take a toll in sweat and loss of moisture in the breath. These liquids must be regularly replaced, preferably by hot drinks, which provide extra calories if they contain sugar.

5-2. RATIONS

Rations provide the needed calories to live and fight effectively. When eaten in their entirety, rations contain the right amount of carbohydrates, fats, protein, and vitamins. The proper intake of these essential items depends on the entire ration being eaten in properly spaced meals. The lack of concern regarding eating caused by the cold, combined with the difficulties and inconvenience of cooking, may tempt soldiers to miss meals. The principles of sound leadership and discipline in cold weather require that meals be prepared and that the entire ration be eaten, warmed when possible. Save snacks (cookies, chocolate bars, etc.) for between meals and when on the march.

5-3. LIQUIDS

Because there is a large amount of fluid loss in the cold, you should drink a minimum of 3.5 quarts of water a day when heavy physical activity is involved. Water is usually available either from streams or lakes or by melting snow or ice. A limiting factor may be fuel needed to melt the snow or ice. To save time and fuel, water should be obtained from running streams or a lake. The milky water of glacial streams must be allowed to stand until the sediment settles. When a hole is cut into the ice to get water, it should be covered by a snow block or something similar to slow refreezing.

Ensure that as much of the daily liquid intake as possible is hot drinks, such as soups and cocoa. Your local command may request a special food allowance of warming and cooling beverages. Main meals should begin with soup, and between-meal snacks should include a hot drink.

When running water is not available, ice or snow must be melted. Ice produces more water in less time than snow. When melting snow, first put a small amount into the cooking pot; then more can be added. Continue this process until there is enough water. It must be purified by boiling rapidly for 15 minutes.

When heating water, use all available stoves, since this is a most time-consuming operation. Melting and boiling enough snow for a drink may take 30 to 40 minutes. Areas designated for ice or snow that is to be used for water must be sited far upwind of the latrine and garbage disposal sites. Before going to bed, ensure that stoves are filled and enough water has been prepared for breakfast. Do not let other soldiers eat snow or ice; doing so lowers the body's core temperature.

Never drink alcohol on operations or when exposed to low temperatures. The aftereffects of alcohol can lead to a feeling of false security. You may forget the main rules to prevent cold injury, like wearing your gloves or not going outside without shoes to relieve yourself.

CHAPTER 6

Hygiene and First Aid

"...the first frostbite cases have made their appearance. We amputated two feet and will amputate many more."

Amosoff

6-1. GENERAL

The care of your body needs special attention in cold weather. Ensuring cleanliness in the field, particularly in the cold, is important. It is easy to let yourself go for long periods without bathing, but if you wish to be part of an effective combat force, you must keep clean.

6-2. HYGIENE TIPS FOR THE SOLDIER

Face, hands, armpits, and crotch should be washed daily. If water is not available, take a "bath" using cornstarch or rub your body briskly with a dry towel. The cornstarch removes excess oil and perspiration from skin and hair.

Shave regularly -- daily if water is available or every 2 to 3 days if water is in short supply. Shaving just before sleeping will allow the face maximum recovery time before going out into the cold, reducing the possibility of superficial frostbite.

Teeth should be cleaned daily. If a toothbrush is not available, a clean piece of cloth wrapped around a finger or a chewed twig may be used as a substitute.

Underwear should be changed at least twice a week. If it is not possible to wash your underwear, crumple it, shake it, and air it for at least 2 hours.

Ensure that all eating and cooking utensils are cleaned after each use.

In the dry arctic you will find the need to oil your unprotected and exposed skin once every week (with an oil such as suntan oil).

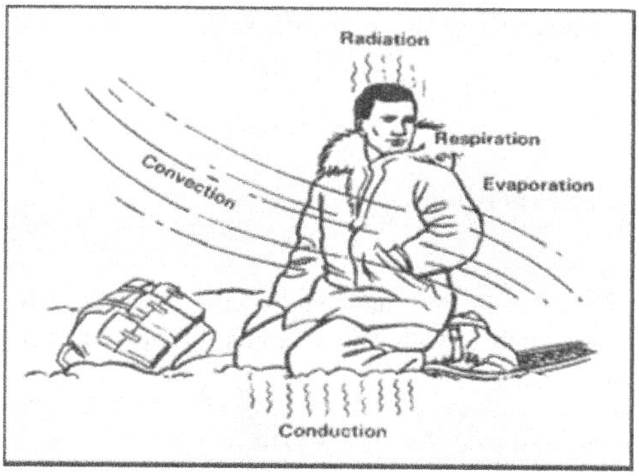

Figure 6-1. Mechanisms of heat loss from the body.

6-3. HYPOTHERMIA

Hypothermia is a lowering of the temperature of the body's inner core. This happens when the body loses heat faster than it can produce it. You must be able to recognize the symptoms of hypothermia and to treat the condition immediately.

PREVENTION

To prevent hypothermia, adhere to the following:

- Stay physically fit.
- Keep active.
- Use the uniform properly and keep it dry.
- Eat properly and often.
- Drink plenty of liquids, at least 3.5 quarts per day, when performing physically demanding work.
- Be prepared for and know how to deal with quick changes in the weather.
- Bivouac early before judgment is weakened by fatigue.

SYMPTOMS

Be aware of the following symptoms:

- Shallow breathing or absence of breathing.
- Faint or undetectable pulse.
- Victim is first cold, then stops shivering.
- Loss of feeling.
- Poor coordination; victim has trouble walking.
- Impaired speech.

- Confusion, withdrawn appearance, depressed, and/or uncoordinated. An uncaring attitude and/or glassy stare are also clues.

TREATMENT

The recommended treatment is as follows:

- Prevent any further heat loss.
- Get the victim out of the wind and into the best shelter available.
- Replace any wet clothing with dry, and place the victim in a sleeping bag, if one is available.
- Place as much insulation as possible between the victim and the ground.
- Add heat by the best available means to the victim's neck, groin, and sides of chest. This can be done by using a hot water bottle, Yukon stove, campfire, or your own body heat.
- Provide the victim with something warm and nutritious to drink if he is conscious. Calories may be added by using sugar in hot, sweet drinks.
- DO NOT massage the victim.
- DO NOT give alcohol to the victim.
- Evacuate the victim to the nearest medical treatment facility as soon as possible. This condition is life threatening until normal body temperature is restored.

6-4. FROSTBITE

Frostbite results when tissues freeze from exposure to temperatures below 32°F. The degree of injury depends upon the wind-chill factor, length of exposure time, and proper protection. Persons with a history of cold injury are prone to frostbite. There may be a tingling, stinging, an aching sensation, or a cramping pain. The skin first turns red and then becomes pale gray or waxy white. Frostbite can be classified as superficial or deep.

WIND CHILL

The combination of wind and low temperatures creates a condition known as wind chill. For example, with the wind calm and a temperature of -20°F there is little danger of wind chill. If the temperature is -20°F and there is a wind of 20 knots, the equivalent chill temperature is -75°F!

Under these conditions there is GREAT DANGER and your exposed skin can freeze within 30 seconds. You also create your own wind when you walk, run, or ski. As examples, walking creates a wind speed of about 5 mph and running 10 mph. Wind is also created by moving vehicles, aircraft propellers, and rotors. The wind-chill chart is provided to show how wind and temperature combine to create wind chill; use it. Cold injury can disable you like an enemy bullet.

WIND SPEED		COOLING POWER OF WIND EXPRESSED AS "EQUIVALENT CHILL TEMPERATURE"																				
KNOTS	MPH	TEMPERATURE (°F)																				
CALM	CALM	40	35	30	25	20	15	10	5	0	-5	-10	-15	-20	-25	-30	-35	-40	-45	-50	-55	-60
		EQUIVALENT CHILL TEMPERATURE																				
3-6	5	35	30	25	20	15	10	5	0	-5	-10	-15	-20	-25	-30	-35	-40	-45	-50	-55	-60	-70
7-10	10	30	20	15	10	5	0	-10	-15	-20	-25	-35	-40	-45	-50	-60	-65	-70	-75	-80	-90	-95
11-15	15	25	15	10	0	-5	-10	-20	-25	-30	-40	-45	-50	-60	-65	-70	-80	-85	-90	-100	-105	-110
16-19	20	20	10	5	0	-10	-15	-25	-30	-35	-45	-50	-60	-65	-75	-80	-85	-95	-100	-110	-115	-120
20-23	25	15	10	0	-5	-15	-20	-30	-35	-45	-50	-60	-65	-75	-80	-90	-95	-105	-110	-120	-125	-135
24-28	30	10	5	0	-10	-20	-25	-30	-40	-50	-55	-65	-70	-80	-85	-95	-100	-110	-115	-125	-130	-140
29-32	35	10	5	-5	-10	-20	-30	-35	-40	-50	-60	-65	-75	-80	-90	-100	-105	-115	-120	-130	-135	-145
33-36	40	10	0	-5	-15	-20	-30	-35	-45	-55	-60	-70	-75	-85	-95	-100	-110	-115	-125	-130	-140	-150
WINDS ABOVE 40 HAVE LITTLE ADDITIONAL EFFECT		LITTLE DANGER						INCREASING DANGER (Flesh may freeze within 1 minute)						GREAT DANGER (Flesh may freeze within 30 secs)								

Figure 6-2. Wind-chill chart.

PREVENTION

Frostbite is preventable if the individual soldier takes care of himself. It seldom occurs in people who are maintaining enough body heat. It is most commonly associated with an overall body heat loss resulting from poor equipment, reduced food intake, dehydration (water loss), overtiring, injury, or a combination of these factors. The feet, hands, ears, and exposed areas of the face are most easily hurt by frostbite and must receive constant care. The following are rules to help prevent frostbite:

- Do not wear tight boots and socks.
- Use the "buddy system" to check exposed areas, especially when wind is present.
- Carry extra socks and mitten liners.
- Do not stay still for long periods.
- Use caution when cold and wind are combined.
- Check feet during halts.
- Ensure that you eat well and have hot drinks as often as possible.

SYMPTOMS OF SUPERFICIAL FROSTBITE

Redness occurs, followed by powdery flaking of the skin. Affected areas of dark-skinned soldiers may appear dull and grayish. Blister formation occurs 24 to 30 hours after exposure, followed by flaking of superficial skin.

SYMPTOMS OF DEEP FROSTBITE

Common symptoms are as follows:

- Loss of feeling occurs in the affected area, leaving it pale yellow and waxy looking.
- When the frostbitten area thaws, it is painful.
- Frozen tissue may feel solid or "wooden" to the touch.
- When exposed to inside temperatures, blisters may appear in 12 to 36 hours.

- Discoloration (red-violet) appears 1 to 5 days after the injury.
- Gangrene usually results.

TREATMENT FOR FROSTBITE

Determine whether the frostbite is superficial or deep. If the exposure time was short, the frostbite will probably be superficial. If the exposure time was long, the frostbite will probably be deep.

The following is a recommended treatment for frostbite:

- Move the casualty to a warm and sheltered area.
- Do not rub the frostbitten area with snow or apply cold water soaks.
- Do not rewarm the affected area by massaging or exposing it to open fire.
- Rewarm the face, nose, or ears by placing your hands on the frozen area. Rewarm frostbitten hands by placing them under clothing and against the body.
- Close the clothing to prevent further loss of body heat.
- Rewarm the feet by removing the boots and socks. Place the bare feet under the clothing and against the abdomen of a buddy. Once the feet are warmed, put on dry socks and boots, if available. If the victim must wear the wet socks and boots, he should exercise his feet by wiggling his toes.
- Loosen tight clothing and remove jewelry.
- Improve circulation by exercise.
- Do not allow the victim to use alcohol or tobacco. Alcohol increases loss of body heat; tobacco causes narrowing of blood vessels in the arms and legs.
- Reassure the victim.
- Protect frozen tissue from further cold or trauma, and evacuate the casualty to the nearest medical treatment facility.

Deep frostbite is most common and harmful to the feet and less common to the hands and ears. When deep frostbite does occur, the victim should be moved to a sheltered area and evacuated to a medical treatment facility immediately. If possible, do not allow the victim to walk if his feet are frozen and avoid treating or thawing the affected area. Thawing of deep frostbite in the field increase pain and invites infection, greater damage, and gangrene. Do not try to thaw deep frostbite; let the physician do it.

6-5. DEHYDRATION

Dehydration can cause serious physical problems and make you more susceptible to other problems, such as frostbite and hypothermia.

PREVENTION

The following are suggestions to prevent dehydration:

- The **minimum** daily liquid requirement for personnel performing demanding physical work in the cold is 3.5 quarts per day.
- By the time you feel thirsty you are already dehydrated. Drink whenever possible, particularly during halts.
- Ensure that canteens are full before any type of movement.
- During periods when water is scarce, do not drink coffee in large quantities. It will pass through you quickly. Water should be the first choice.
- Check your urine spots in the snow. Dark yellow or brown indicates dehydration.

SYMPTOMS

Common symptoms are as follows:

- Red indicates a severe condition needing immediate medical evacuation to the nearest medical treatment facility.
- Lack of appetite.
- Dry mouth, tongue, and throat.
- Stomach cramps and/or vomiting.
- Headaches.

TREATMENT

Recommended treatment is as follows:

- Keep the victim warm.
- Give him plenty of liquids.
- Make him rest.

6-6. CARBON MONOXIDE POISONING

Whenever a stove, fire, gasoline heater, or internal combustion engine is used indoors, there is danger of carbon monoxide poisoning. Fresh air in living and working quarters is vital. Carbon monoxide is a deadly, odorless gas.

PREVENTION

The following are suggestions to prevent carbon monoxide poisoning:

- Use stoves and lanterns in well-ventilated areas and tents.
- Ensure that stoves and lanterns are functioning properly.
- Do not let personnel warm themselves by engine exhaust.
- Always have windows cracked in vehicles with a heater in use.

- Use a tent guard or shut the stove off when sleeping.

SYMPTOMS

Common symptoms are as follows:

- Headache, dizziness, confusion, yawning, weariness, nausea, and ringing in the ears.
- Bright red color on lips and skin.
- Victim may become drowsy and collapse suddenly.
- If personnel are found unconscious in an enclosed shelter, carbon monoxide poisoning should be suspected.

TREATMENT

Adhere to the following for proper treatment:

- Move the victim to open air.
- Keep the victim still and warm.
- If the victim is not breathing, administer mouth-to-mouth resuscitation.
- Administer cardiopulmonary resuscitation, if necessary.
- Immediately evacuate the victim to the nearest medical treatment facility.

6-7. SNOW BLINDNESS

Snow blindness is caused by ultraviolet and ultrablue rays of the sun being reflected from a snow-covered surface into the eyes. This condition can occur even in cloudy weather. In fact, it is more likely to occur in hazy, cloudy weather than in sunny weather.

Figure 6-3. Snow blindness.

PREVENTION

Prevention is easy; wear your issued sunglasses. If sunglasses are not available, make slitted glasses from cardboard, thin wood, tree bark, or similar material.

SYMPTOMS

Common symptoms are as follows:

- A scratchy, sandy feeling under your eyelids.
- Redness and watering of the eyes.
- A headache.

TREATMENT

Adhere to the following for proper treatment:

- Blindfold the patient using a dark cloth.
- Reassure the patient.
- Evacuate the patient to a medical treatment facility.

6-8. SUNBURN

Sunburn is a serious hazard because of the reflective qualities of the snow, especially at higher elevations.

PREVENTION

Require soldiers to use a sunscreen and lip balm.

SYMPTOMS

Redness of skin with slight swelling is one symptom of sunburn. Prolonged exposure to the sun may cause pain and blistering and, in severe cases, chills, fever, and headaches.

TREATMENT

Soothing skin creams may be helpful if the swelling is not severe.

6-9. TENT EYE

Tent eye is an inflammation of the eye caused by fumes from stoves and lanterns used in a poorly ventilated tent or shelter. It is prevented by proper ventilation of tents or shelters. The treatment is fresh air.

6-10. TRENCH FOOT/IMMERSION FOOT

Trench foot and immersion foot are injuries caused by the prolonged exposure of skin to cold or wet conditions. In extreme cases, the skin dies and amputation of the foot or leg may be necessary.

PREVENTION

The following helps prevent trench foot or immersion foot:

- Change to dry socks at least twice daily.
- Dry and massage your feet regularly with foot powder.
- Wear VB boots and change socks regularly.
- If leather boots are worn, dry them whenever possible.

SYMPTOMS

In early stages, the feet and toes are cold, numb, and still, and walking becomes challenging. The feet swell and become painful.

TREATMENT

In early stages, keep the feet dry, clean, and exposed to the air. In later stages, evacuate the victim to a medical facility.

6-11. CONSTIPATION

At very cold temperatures, personnel tend to put off the natural urge to have a bowel movement, resulting in constipation. This can be a painful casualty producer.

PREVENTION

Eat fresh or canned fruit and eat regularly. Drink plenty of liquids. Try to have a daily bowel movement at your normal time.

SYMPTOMS

Some common symptoms are stomach cramps, dizziness, and headaches.

TREATMENT

Prevention is the best treatment. If symptoms persist, get medical help.

6-12. DIARRHEA

Diarrhea can be treated by drinking tea leaves with water.

6-13. HEAT CRAMPS

Heat cramps are caused by an excessive loss of salt in the body. This loss of salt causes muscles to spasm, and they are unable to relax. After prolonged physical activity in a cold-wet environment, the signs and symptoms of heat cramps could appear in the arms, legs, and/or stomach. It is possible to experience heat cramps in cold-weather environment when undergoing extreme physical exertion.

SYMPTOMS

The soldier experiences muscle cramps in his arms, legs, and/or stomach. He is pale and has wet skin and is experiencing dizziness and extreme thirst.

TREATMENT

Have the soldier drink one fourth of a canteen of water. Give the soldier one canteen of water with one packet of salt stirred in. Have him drink the canteen of salted water over a 30-minute period. If cramping of muscles continues, evacuate the soldier to the nearest medical treatment facility.

> **NOTE**
>
> **Do not give the saltwater to a soldier who has an upset stomach. Have him drink a canten of unsalted water.**

6-14. HEAT EXHAUSTION

Heat exhaustion is caused by dehydration and loss of body salt from extreme physical activity in a cold-wet environment. It is possible to experience heat exhaustion in a cold-weather environment when undergoing extreme physical exertion.

SYMPTOMS

The soldier may feel dizzy, weak, and/or faint. The skin feels cool and moist to the touch. He may feel nauseated or have a headache.

TREATMENT

The following is a recommended treatment for heat exhaustion:

- Have the soldier drink one canteen of water to relieve the symptoms.
- If he complains of cramps, give him one canteen of water with one packet of salt stirred in. Have him drink the salt solution over a 30-minute period.

> **NOTE**
>
> **Do not administer salt if the soldier has an upset stomach. Have him drink a canteen of unsalted water.**

- Loosen all tight-fitting clothes.
- Elevate the feet above the level of the heart.
- If the soldier is unable to drink water due to an upset stomach or if the symptoms have not improved within 20 minutes, evacuate the soldier to the nearest medical treatment facility.

CHAPTER 7

Bivouac Routine

> ***"Had we lived, I should have a tale to tell of the hardihood, the endurance and the courage of my companions which would have stirred the hearts of every Englishman. These rough notes and our dead bodies must tell the story."***
>
> ***CPT Robert Scott***

7-1. GENERAL

The essential requirements for survival and mission accomplishment in the cold/arctic environment are warmth, food, and shelter. As covered in Chapter 3, you are provided with tents that can be erected quickly. Tents can be lost or destroyed and separated from the ahkio groups in the confusion of combat. When this occurs, timber and snow, which are usually available, will be used for shelters. A detailed explanation of the bivouac routine should be contained in the unit SOP.

7-2. BIVOUAC SITE

Bivouac sites, other than those that are permanent nature in rear areas, either are tactical or are designated to provide living accommodations within a defended position. The tactical bivouac is usually not occupied for more than one night, is established at or after last light, and needs little preparation. It is organized for a rapid and disciplined response to enemy action. The defensive position is likely to be occupied for longer periods and must be established accordingly. The layout and placement of tents and shelters must be coordinated within the defense and in accordance with the unit SOP.

7-3. SITE SELECTION

The selection of bivouac areas is important and needs careful consideration. The challenge of area selection varies with the tactical situation, weather conditions, and terrain. The bivouac area will be tactically located in accordance with the principles of security and defense.

There must be concealment from air and ground observation and wind. Forested areas pose few challenges in comparison to areas that are located above the tree line. In selecting areas, ensure that local camouflage materials are available for use with the snow camouflage nets. Areas that offer better bivouac sites than others are explained below.

FORESTED AREAS

Forests provide excellent sites. Camouflage is easy and building materials, firewood, and wood for insulation are readily available. Although forests provide the best bivouac site, temperatures there will normally be colder than those in open areas.

MARSHY GROUND

In winter, good sites are frozen swampy areas, the banks of streams, and the shores of lakes. Some swampy areas may not freeze because of warm springs or marsh gases. Deep valleys should be avoided since colder air gathers there.

OPEN AREAS

Avoid open areas because of strong winds, drifting snow, and poor concealment. If avoidance is not possible, then pitch tents where they can be sheltered by natural windbreaks. These are normally found in depressions or behind ridges of ice on lakes. In areas where natural wind falls do not exist, build snow walls to provide wind protection. In areas with high winds, snow gathers on the lee side, making it necessary to clear the side and tops of the tents, often to prevent the weight of the drifting snow from collapsing the tent.

MOUNTAINOUS AREAS

Mountainous terrain is characterized by strong winds, cold, and a general lack of concealment above the timberline. The wind over head creates extensive lees near cliffs and steep rock faces. Avalanches can occur on steep slopes; avoid their likely path. Signs to look for are:

- Heavy snow overlaying a crust of ice (usually the result of a freeze at the end of a period of thaw).
- New snow on steep slopes if the temperature is rising.
- Evidence of avalanches in previous years; an area cut through the trees (branches stripped off, and/or broken tree stumps at different heights) or a smooth clear strip of ground down the mountainside.

Figure 7-1. Location of tent.

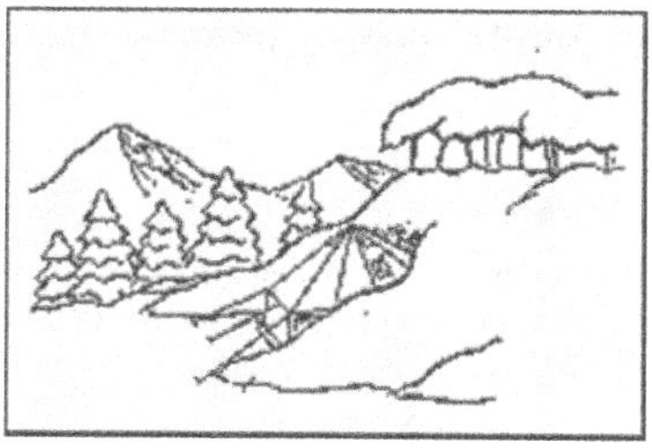

Figure 7-1. Location of tent (continued).

When contact with the enemy is likely, the unit SOP will identify specific considerations when selecting the site for a tactical bivouac. For example:

- The location of the enemy.
- Enemy patrol habits.
- The air threat.
- Dominating features that identify enemy avenues of approach.
- The wind direction that can carry sounds of movement (the bivouac site should be downwind of likely enemy avenues of approach, especially when wind speeds are low).
- Availability of natural camouflage.
- Amount of moonlight and weather conditions. In good weather or bright moonlight, the bivouac site must be in an area of deep shadow (in the middle of the woods or on the northern slope of a ridge) or the tents will be silhouetted.

7-4. RECONNAISSANCE OF BIVOUAC SITE

The main body should be preceded by a reconnaissance party whose task is to lay out the bivouac site. They should:

- Identify the bivouac site and a false position.
- Establish a trail plan.
- Select defensive positions.
- Mark the location of tent sites.
- Identify areas for cutting boughs.
- Act as guides for the main body on its arrival.
- Provide security during occupation by the main body.

7-5. TRAIL PLAN

Trail plans must be established before entering the bivouac site. Normally, the incoming trail will be extended well beyond the area of the bivouac and may lead to a false position. Leading off the incoming trail establish another trail, usually at an extreme angle to entry. The trail network that leads to each subordinate unit area, and within, must follow the inside of the tree line. This prevents enemy observation of trails from the air or ground.

7-6. OCCUPATION

Guides from the reconnaissance party will arrange to meet the main body and identify the layout of the bivouac site. Once the unit is in position, the unit should:

- Identify a temporary area for weapons and equipment during establishment of the site.
- Identify the exact location and type of defensive positions to be built.
- Determine exact tent sites regarding defensive positions, natural shelter, and camouflage.
- Identify sleeping areas and type of shelters to be built when tents are not used.

7-7. BOUGH-CUTTING AREAS

Bough-cutting areas for bedding or improvised shelters should be for all personnel in the unit. These areas must be where activity is not likely — well behind the unit's defensive area or in a decoy bivouac site.

Cut boughs from the bottom portion of trees above the snow and leave the trees standing. Be sure to cut only a few boughs from each tree. A tree stripped of its limbs is easily observed from the ground and from the air.

Cut firewood near the bough-cutting area to allow the same trail to be used. Dead spruce, pine, or squaw wood makes the best firewood. Standing deadwood burns the best and hardwoods the longest. In an arctic and subarctic environment, true hardwoods are rare. Of the trees available, birch, tamarack, and spruce are the hardest woods. Burn the upper parts of dead trees during daylight because they give off lighter colored smoke. The lower parts of the trees contain resin and tar, which burn well but make more and darker smoke. In the daylight this will allow your position to be easily identified by air and ground observation.

7-8. GARBAGE

All garbage should be double bagged and transported to the rear or disposed of in accordance with the unit SOP. Units should never leave evidence of their presence. Carry all trash, if necessary, until it can be disposed of.

7-9. LATRINE

Normally there will be a central latrine within the bivouac site, if the site is not too widely spread. One latrine can usually serve the needs of a platoon-size element. It must be placed downwind of the bivouac site, but not so far that soldiers won't use it. The area identified can be as elaborate as a pit or cross-cut-type latrine that is made windproof by branches, snowblocks, and ponchos. It should be properly camouflaged. It can also be an area where plastic bags are located. Double-bag, tie, and transport all waste to the rear for disposal as identified in the unit SOP. Site urinal areas around a "dedicated tree" or the gasoline tripod, which will hide the discolored snow and will be easily spread about when leaving the bivouac site.

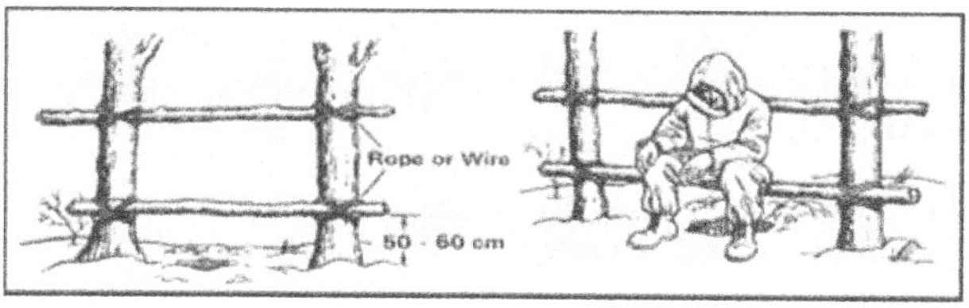

Figure 7-2. Cross-tree type latrine.

7-10. WATER POINTS AND SNOW AREAS

When obtaining water for drinking purposes by melting snow, set aside and restrict an area for this purpose only. The site must be upwind from the bivouac area and isolated from the latrine and garbage areas. The snow must be clean, white, and free of fungus from trees. NEVER use snow that is discolored. Remember — disinfect the water using purification tablets and bring it to a rolling boil for a minimum of 15 minutes.

7-11. STORAGE

Storage problems in the winter are increased by snow, low temperatures, thaws, and limited storage space. Space in any shelter is limited. Store only those items inside the tent that are affected by the cold, or must be immediately available. The remaining equipment should be centralized, well marked, and covered. Store ammunition and fuel separately. Place wood or metal containers on boughs or poles to keep the containers from freezing to the ground. The unit SOP will specify what equipment is placed inside or outside of the tent and where.

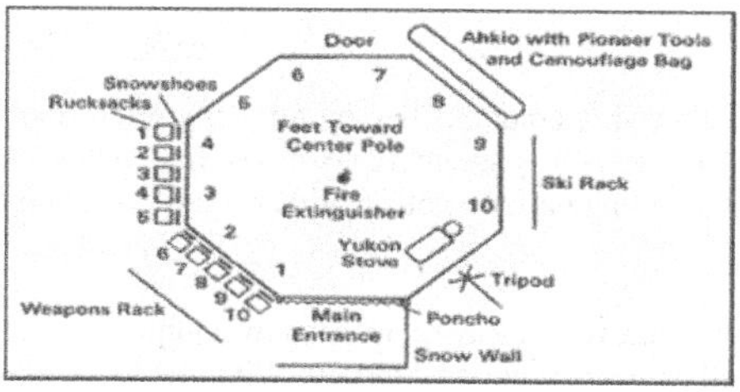

Figure 7-3. Ten-man tent.

CHAPTER 8

Individual Movement

"Anticipate every situation which can arise in battle and think out your reaction to it well in ADVANCE."

Hoote

8-1. GENERAL

Snow affects movement many ways. Heavy snow cover hinders or slows movement, either cross-country or on roads. Snow cover also blankets many terrain features and hides obstacles.

The effect of snow on movement varies greatly with its depth and physical characteristics. Men on foot cannot travel well over flat terrain or roads when the snow is deeper than 30 centimeters. However, movement can be improved by the use of skis or snowshoes.

SKIS	SNOWSHOES
• Faster • Less Work • Harder to Learn	• Slower • More Work • Easy to Learn

Figure 8-1. Using skis or snowshoes.

8-2. SKIS

Skis provide greater speed in moving, particularly over cleared trails. Troops mounted on skis and towed behind small unit support vehicles, or similar vehicles (ski-joring), are effective means for rapid cross-country movement. In heavy vegetation, skiing will always be more physically taxing than snowshoeing.

8-3. SNOWSHOES

Learning how to use snowshoes requires less training. Snowshoe movement is more practical in confined areas, such as assembly positions, field trains, and mortar and artillery positions. If traveling on foot, you should become skilled on both skis and snowshoes.

Your ability to use, and the availability of, over-snow equipment is a major factor in learning to travel well over snow-covered terrain. With previous training in cross-country skiing and if equipment is available, travel on skis is recommended. In most snow conditions and over most types of terrain, skis provide the fastest and most energy-saving form of travel. Use of snowshoes requires little training, but your speed will be slower and travel more exhausting.

Snowshoes are **more** exhausting than skis. However, a light crust on the surface of the snow prevents skis from sinking and provides for fast and easy skiing. A crust hard enough to support a man makes travel on foot feasible, but even then travel on skis is recommended.

In a survival situation, improvise equipment for travel if snow is loose and deep. Make snowshoes of willow or other springy wood using a wood separator and thong, wire, cord, etc.

USE SKIS WHEN:

- Speed is essential.
- Long distance must be traveled.
- Stealth is necessary.
- Conditions allow.

USE SNOW SHOES WHEN:

- Moving through heavy brush.
- Speed is not essential.
- Troops are not proficient on skis.

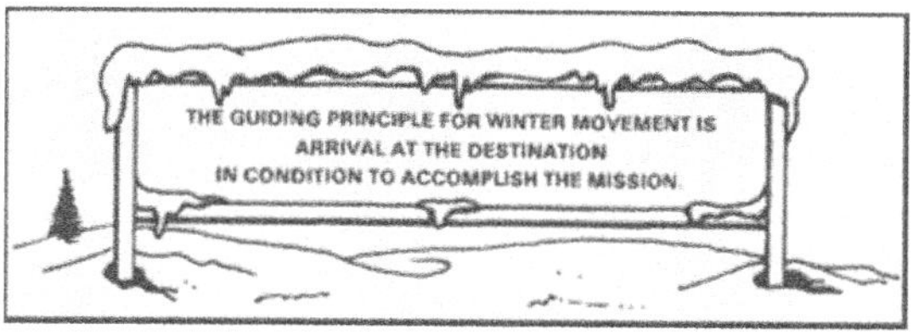

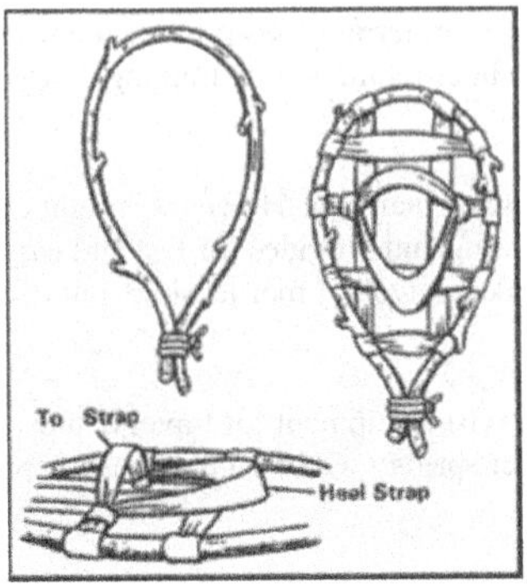

Figure 8-2. Improvised snowshoes.

CHAPTER 9

Land Navigation

"No - I've never been lost; but for about three weeks I was kinda puzzled."

Brigadier General Fergusson

9-1. GENERAL

In some cold-weather environments, accurate up-to-date maps may not be available and landmarks may be few or nonexistent. When the terrain is snow covered, one piece of ground looks like the other. Whatever the circumstances, you must be able to find your way — if not by map reading, by some other means of navigation.

9-2. CHALLENGE OF LAND NAVIGATION

Only poor road nets may exist in some areas. These will be of little use in cross-country navigation. If large-scale maps exist, they may be unreliable.

Long nights, falling or blowing snow, and fog limit visibility in winter. Tracks and landmarks may be erased by heavy snowfall. Accurately identify landmarks, through both ground and air observation, to prevent mistakes.

Conditions of daylight are different from those found in temperate regions (moderate climate). In winter there are long nights and short days.

Magnetic disturbances may be common, thereby making compass readings unreliable. Magnetic declinations may be extreme and can vary considerably at close points.

Many small lakes, ponds, and creeks will be found that are not on the map. The size and location of waterways may vary from year to year and from season to season and cannot be relied on to pinpoint a position. Large lakes and rivers can be used as important aids to navigation and movement.

9-4. DISTANCE AND DIRECTION

When traveling from one point to another, it is necessary for you to know the distance between them beforehand, then plot your journey accordingly.

9-5. MEASURING DISTANCES

PACING

In temperate zones, pacing is the simplest method of measuring distance. In cold-weather areas, snow cover and the use of skis and snowshoes make pacing challenging, and it must be practiced repeatedly.

USE OF FIELD WIRE

The best method for accurate ground measurement is for two men to use a piece of salvage field wire 50 meters long. The lead man marches off in the desired direction trailing the wire behind him. The second man signals by jerking the wire when the first has gone 50 meters and then moves forward to the first man's position. The first man marks the spot and continues forward for another 50 meters. This is repeated until the desired distance is reached.

ESTIMATION

If no other method of measurement is available, the length of time spent in steady progress along a route may be used to estimate the distance traveled. The accuracy of this method depends on how well you know your marching speed over all types of terrain on foot, skis, or snowshoes.

9-6. DETERMINING DIRECTION

COMPASS

The use of the compass is the most common way to determine direction. The magnetic field becomes weaker as you get nearer to the magnetic poles and can be easily affected by other objects. It is important to make sure that the compass is not affected by iron fragments in the ground, vehicles, or metal objects on you.

THE SUN

There are two ways to determine direction by the sun:

- At noon, in the northern hemisphere, shadows fall to the north of objects; in the southern hemisphere, to the south of objects.
- Estimate direction by using a watch and the sun. In northern latitudes the hour hand is pointed toward the sun. An imaginary line halfway between the hour hand and 1200 will point south. In southern latitudes the procedure is different. Then point the 1200 toward the sun. An imaginary line halfway between the hour hand and 1200 will point north.

NATURE

Nature will help in determining direction as in the following samples:

- Single evergreen trees, growing away from other trees, will always be more bushy on the south side.
- The bark of birch and poplar trees is whitest on the south side and darkest on the north.
- You should know from which direction the prevailing winds blow. Trees will be bent in the direction the wind is blowing or expected to blow.

THE STARS

The North Star may be used to determine true north. This star can be easily found by its position relative to the Big Dipper. In latitudes under 70 degrees when traveling north, the North Star makes a good steering mark since its bearing is usually only 1 degree from true north and is never more than 2 1/2 degrees away. In higher latitudes above 70 degrees, the North Star is too high in the sky to indicate good direction.

A star near the horizon can be used with a bearing within 2 degrees of your compass course. However, it will only be good for up to 30 minutes when you are traveling north and only 15 minutes when traveling south. After these intervals, select another star.

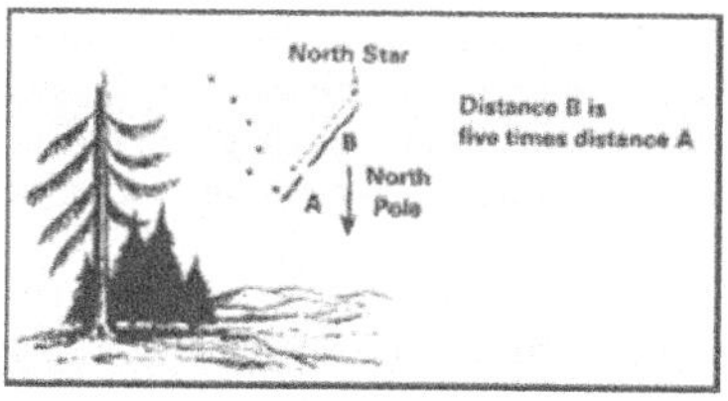

Figure 9-1. Determining direction using North Star.

South of the equator, the constellation Southern Cross will help you locate the general direction of south and any other direction. This group of four bright stars is shaped like a cross that is tilted to one side. The two stars forming the long axis, or stem, of the cross are called pointers. From the foot of the cross, extend the stem five times its length to an imaginary point. This point is in the general direction of south. From this point, look straight down to the horizon and select a steering point.

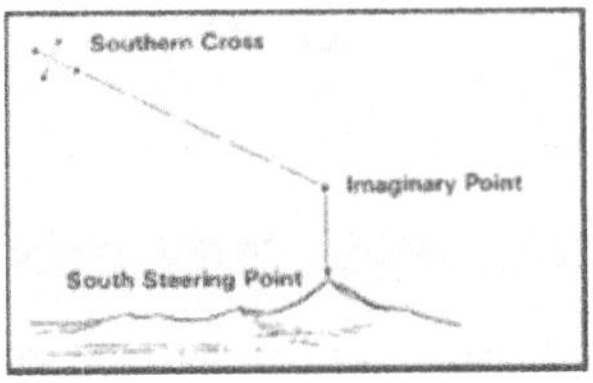

Figure 9-2. Determining direction using Southern Cross.

9-7. NAVIGATION BY DEAD RECKONING

Dead reckoning consists of plotting and recording a series of courses from a known point, each measured for distance and direction. These courses lead from the starting point to the destination. This lets the navigator determine his position at any time, either by following his plot or by comparing his actual position on the ground to his plotted course. All the details are previously estimated and recorded.

EQUIPMENT FOR DEAD RECKONING

- Maps are used for choosing the route and for plotting the route taken as the march progresses.
- Aerial photographs are usually used in addition to maps. They must be vertical photos taken during the same season in which they are to be used.
- The compass is used as the main direction-finding instrument.
- The protractor is necessary in dead reckoning for plotting direction and distance on the map, aerial photo, or chart. When using it, remember to convert magnetic azimuth to grid azimuth.
- A route card is used to outline the plan of the proposed march. Keeping a log is mandatory. It is used to record the distance and direction.
- The log is a written record of where the soldier has started, the distance and direction traveled, and his position at any time in relation to the planned route, destination, and starting point.

THE DRILL OF DEAD RECKONING

Know the starting point and destination and, having examined the map or aerial photo for the best route, prepare a route card. This describes each leg of the proposed route in terms of distance and direction. As the march continues, keep a careful record of each bearing taken and the distance, and record this in the log. There may be a change to the planned route because of terrain or enemy action. Therefore, the necessary changes are made using the map or aerial photo as a guide, and they are recorded in the log.

9-8. HINTS FOR THE NAVIGATOR

The navigator must keep the compass warm to speed up recording his position. When no steering marks exist to the front, he should march on a back azimuth. A steering mark may be a natural or man-made structure to the rear. If the steering mark is not clear, he must keep an eye on it after taking an azimuth so that it doesn't become lost. Never take magnetic azimuths in the area of metal objects.

When visibility is poor, only close-in steering marks will be available. Under these conditions try to pick up steering marks along the correct azimuth. This can be done correctly only when the route to each steering mark follows a straight line. Take compass checks often to ensure that the correct azimuth is being followed. Set the compass for a night course. A steady unshifting wind will also aid in keeping direction.

Depending on the area of the world, other signs can be used. For example, in the barren and semibarren lands of eastern Canada, snowdrifts point to the northwest and windblown trees to the southeast because of prevailing winds. In Scandinavia, scratches on rocks and cliffs caused by the glacial period (ice age) run northwest to southeast.

CHAPTER 10

Survival

"Each day the number of my dogs dwindled and the sleds were broken up to cook those animals that we ate ourselves."

Peary, Arctic Expedition, 1906

10-1. GENERAL

Your mission may take you to an arctic or subarctic region. You expect to have all personal equipment and other members of your unit with you wherever you go. But you could find yourself alone in a remote area with little or no personal gear.

10-2. ACTION WHEN LOST

Brief the soldier on the route before setting out on a march; he should memorize details if possible. A soldier can become separated from the main body of troops, especially if an action is fought. This applies mainly to night battle or bad weather. It can also happen if he is careless and march discipline is poor.

WHEN LOST WITHIN A KNOWN LOCALITY

If all is quiet and there are no battle sounds or aircraft to guide you to friendly lines, go back to the last known position. If you cannot go back, consider carefully the journey already made and the distances and directions from the starting point.

Opinions should be taken from the group. Search parties should be sent out to find the next known steering mark. Trails must be marked carefully so that the search party can return to guide the main group forward or to rejoin the group, should their search be unproductive. Meanwhile, the rest of the group should seek shelter.

If the search party fails to find any known location, perform the steps below, but remember to think clearly and keep calm.

CONDUCT WHEN LOST

When you are certain that you are lost:

- Stay together. Meet your emergency as a group; don't wander off as individuals; never allow stragglers.
- Take shelter. If you have a tent, set it up and get in it. If you do not have a tent, build one of the improvised shelters. The type of shelter is not important as long as you protect yourself from the weather.

- Keep warm. If you are lost in a forest, light a fire (depending on the enemy situation). If you are lost in an area where there are no trees, do not use your gasoline stove to heat your shelter. Use your stove only to prepare hot food and drinks. Conserve your fuel supply; you won't get any more. Get into your sleeping bag to stay warm.
- Check food supply. All the food and water in the group should be collected and rationed systematically. Supplement the food supply by shooting large animals and trapping or snaring small ones.
- Check communications. If you have a radio set, use it to contact your parent unit for help. If you are in a forest, prepare fires to attract search planes. Smoke is better by day and flame at night. Mirrors, ground signals, etc., are helpful in attracting search parties. Arrange necessary ground/air signals for search planes.
- Rest. The more you rest the less energy you expend and the less food you need.
- Adhere to survival tips. Carry safety matches inside the butt plate of your rifle or in your pockets. Carry candles, which are easy to light, provide heat and light, and help in starting a fire. If you have no matches, fire may be produced by striking the flint on your waterproof match box with the blade of a knife and collecting sparks in dry tinder. Dry moss and bark make good tinder to start your fire. Fires should be built in layers with sticks being laid out like the spokes of a wheel.
- Carry a two-cell flashlight. It can be seen from the air or on the ground as far as line-of-sight will permit.
- Ground/air signals can be made with small fires to form the identification characters (see Appendix C).
- Survival depends on clear thinking and resourcefulness. The main things to remember are to keep calm, think, try to help each other, keep together, and keep warm.

10-3. IMPROVISED SHELTERS

Sometimes tents or other regular shelters are not available. In summer, if the weather is mild, you may need protection only from insects. In the winter, however, you cannot stay in the open for long periods unless you are moving. In case of emergency, you must know how to protect yourself from the effects of the weather.

If natural shelters such as caves or rock ledges are available, they should be used. If they are not, a temporary improvised shelter must be built.

The type of shelter to be built depends on the equipment and materials available. By the proper use of materials available, some sort of shelter can be built during any season of the year. In open terrain a shelter can be built using ponchos, canvas, snow blocks, or other materials. Snow caves, snow trenches, snow houses, or snow holes may be built in the winter. In the woods, a lean-to is normally preferred to other types of shelter. In cold-weather areas, nature provides you with the means to prepare a shelter. Your comfort, however, greatly depends on your initiative and skill at improvising.

PONCHO SHELTERS

The simplest type of shelter can be made by pulling the poncho over your sleeping bag. For additional comfort, various types of shelters and lean-tos can be made by attaching your poncho to trees, tree branches, or poles.

To build a one-man shelter from one poncho, spread the poncho on the ground, hood side up. The hood opening must be tightly closed by adjusting and tying the drawstrings. The poncho is raised in the middle of its short dimension to form a ridge, with the stake out at the corners and sides. Snow, sod, or branches are used to seal the side and one end of the shelter to provide additional protection from the wind and to retain heat.

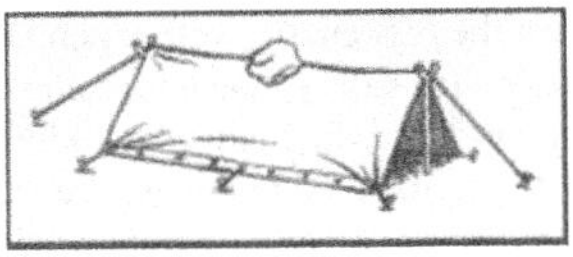

Figure 10-1. One-man shelter.

A two man shelter can be made from two ponchos. Spread the ponchos on the ground with the hood side up and the long sides together so that the snap fastener studs of one poncho may be snapped into those of the other poncho, The ponchos are raised where they are joined to form a ridge. The shelter is then staked out at the sides and corners. A third poncho may be snapped into the other ponchos to form a ground cloth.

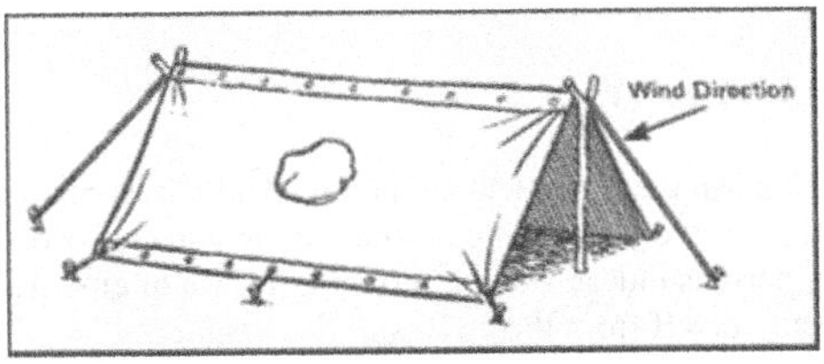

Figure 10-2. Two-man shelter.

LEAN-TO SHELTER

The lean-to shelter is built in forested areas and is made of trees and tree limbs. A poncho, a piece of canvas, or a parachute, in addition to boughs, may be used for covering. Depending on the number of people to be sheltered, two types of lean-tos (single and double) are built. Boughs should be laid in the shelter like overlapping shingles (the same as on a live tree).

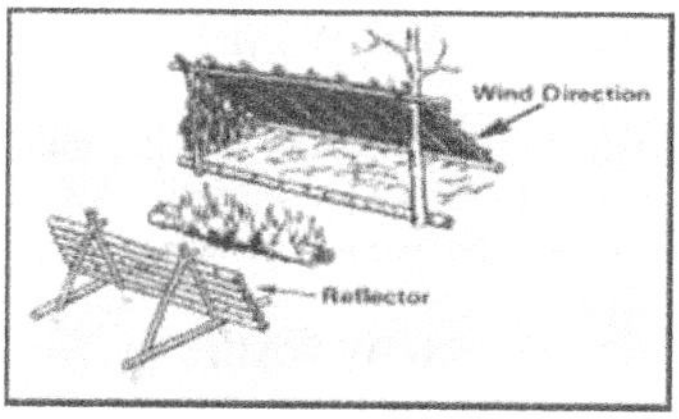

Figure 10-3. Single lean-to.

To save time and energy, use two trees the right distance apart and sturdy enough to support the crosspiece. If suitable trees are not available, use two forked poles or build two A-frames to hold the crosspiece. A large log is laid at the back for added height. Stringers about 3 meters long and 10 centimeters in diameter are then placed 45 centimeters apart from the crosspiece over the top of the log in the rear. Available coverings, ponchos, boughs, and so forth are then placed on top of the stringers. A double lean-to is made by building two single lean-tos facing each other with the fire in between.

Any kind of an open fire can be used to heat the lean-to. The best kind for a large lean-to is the log fire. Precautions must be taken to prevent the fire from burning too hot and burning down the shelter. A fire reflector may be built of green logs or poles to reflect the heat into the shelter and to serve as a windbreak to keep the fire burning steadily.

TREE SHELTERS

In wooded areas, the tree-pit shelter furnishes temporary protection. Select a large tree with thick lower branches and surrounded with deep snow. Enlarge the natural pit around the tree. The walls and floor are lined with branches and boughs. Try not to disturb the snow that may be on the branches.

An emergency shelter for one man can be built by cutting down a coniferous tree at a point about 1 meter from the ground.

Figure 10-4. Tree-pit shelter.

After the tree has fallen, the branches on the underside are trimmed away and the cut material used for insulation on the ground.

WIGWAM

A wigwam or tepee can be built in wooded areas by tying a number of poles near the top and spreading them at the bottom to form a large circle. This framework is then covered with ponchos, a parachute, or any other suitable material.

SNOW WALL

In open terrain with snow and ice, a snow wall may be built for protection from strong winds. Blocks of compact snow and ice are used to build a windbreak. Be careful not to get covered up by drifting snow, which may accumulate on the lee side of the wall.

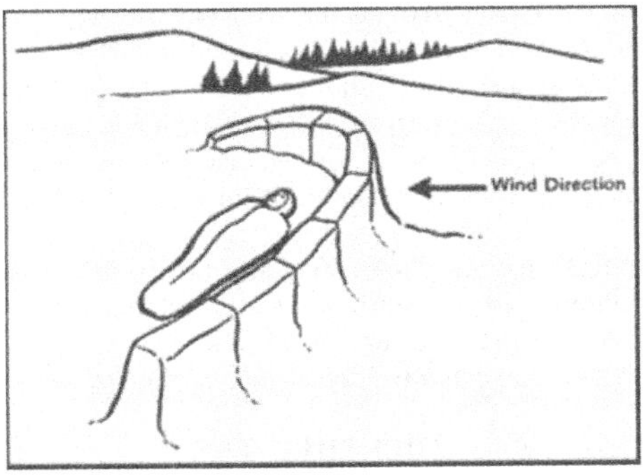

Figure 10-5. Snow wall.

SNOW TRENCH

A snow trench provides shelter quickly. It is built by burrowing into a snowdrift or by digging a trench in the snow and making a roof of ponchos or ice and snow blocks supported by skis, ski poles, or snowshoes. Boughs can be used for covering the roof and for the bed. Figure 10-6 shows one method of roofing over the trench by leaning snow blocks against each other.

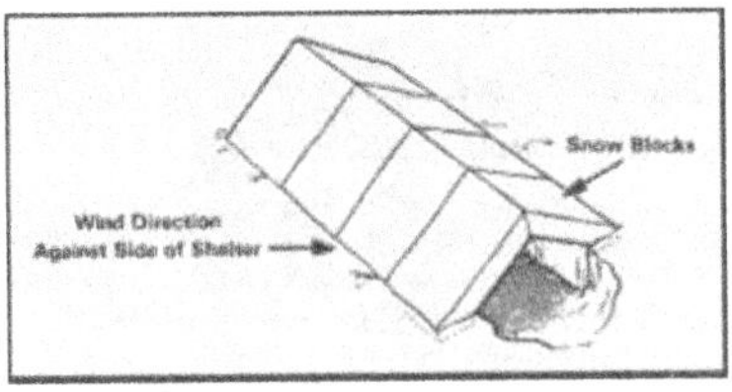

Figure 10-6. Snow trench.

SNOW CAVES

A snow cave can be used as an improvised shelter in the open areas where deep and compacted snow is available. Normally, a suitable site can be located on the lee side of a steep ridge or riverbank where drifted snow collects, and it is wind packed.

The tunnel entrance must give access to the lowest level of the chamber. This is the bottom of the pit, where cooking is done and equipment is stored. The snow cave must be high enough to provide comfortable sitting space. The sleeping area must be on a higher level than the highest point of the tunnel entrance, where there is warmer air.

The roof must be arched for strength so that drops of water forming on the inside will not fall on the floor. This water will follow along the curvature of the roof and sides, glazing over the walls when frozen. The roof must be at least 30 centimeters thick. The size of the snow cave depends upon the number of men that will occupy it. A large cave is usually warmer and more practical to build and maintain than several small caves.

The shape of the snow cave can be varied to suit conditions. When the main cave is built, short side tunnels are dug to make one- or two-man sleeping rooms, storage space, latrine, or kitchen space.

In building the snow cave, locate a deep snowdrift at least 3 meters deep. Newly fallen, powdery, or loose snow should not be used. The depth of a snowdrift may be tested with a sharpened branch or young tree about 4 meters long, or with the ski or shorter ski pole. The entrance must be chosen carefully so that wind will not blow into the cave or block the entrance by drifting snow. The entrance should be built so that it is about 45 degrees from the downwind side. A small tunnel is burrowed directly into the side of the snowdrift for 1 meter. A chamber is excavated from this tunnel. Excavation is done to the right and left, so that the length of the chamber is at right angles to the tunnel entrance. Personnel doing the digging will become wet from perspiration and from the snow inside the cave. They should wear the minimum amount of clothing to ensure that they have a change of dry clothing when finished.

The cave can be heated with the one-burner gasoline stove or with candles. The fires should be put out when personnel inside the cave are sleeping, thus reducing the danger of fire, asphyxiation, or carbon monoxide poisoning. If the weather is severe and it is necessary to have a fire while sleeping, an alert fire guard must be posted in each cave. The ventilation holes must be inspected every 2 or 3 hours to ensure that they have not become clogged by snow or ice.

To ensure that the cave is warm, the entrance should be blocked with a rucksack, poncho, or snow block when not in use. All available material such as ponchos, cardboard from ration boxes, brush, and boughs should be used for ground insulation.

Walking on the roof may cause it to collapse. At least two ventilator holes are used, one in the roof and one in the door. A ski pole can be stuck through the roof ventilator to clear it from the inside. Extra care must be taken to keep air in the cave fresh when cooking or heating. The entrance should be marked by placing a pair of skis, or other equipment, upright on each side of the entry way.

Building a snow cave requires practice. Large shelters are usually more efficient for squad-size units and larger units. The most critical decision is when and where to stop and build the shelter.

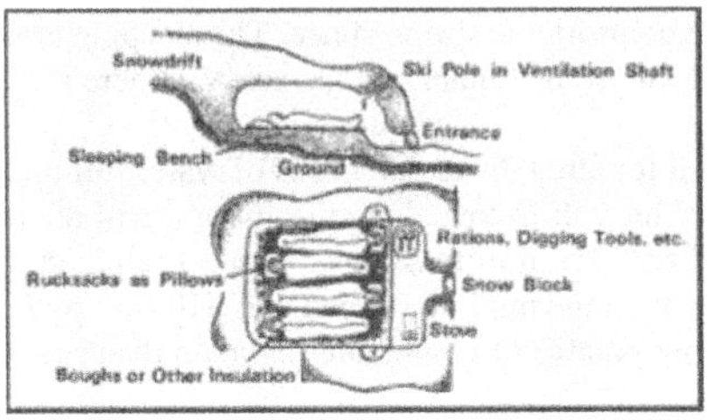

Figure 10-7. Snow cave.

INDIVIDUAL SNOW HOUSE

If you are alone and have need of a shelter quickly, you can build a small individual snow house. It is box-shaped, and the only building materials needed are snow blocks. Cut snow blocks 40 by 40 by 80 centimeters. Lay them on edge to a length of 2.5 meters and a width of 1 meter. Build the walls up to a height of about 1 meter. When you have the walls built, make an A-shaped roof by leaning together pairs of blocks. Fill in the ends with snow blocks and chink all the crevices with soft snow. Cut a hole in the end of the leeward side large enough to crawl into. When you are inside, make one or two small holes near the top for ventilation. Block up the end with a snow block or with your rucksack.

Figure 10-8. Individual snow house.

SNOW HOUSE OR IGLOO

An igloo makes an excellent shelter in snow and glacier areas when other materials are not available or when the snow is so hard that it can be cut into blocks. Cracks in the igloo are then covered with loose snow. To increase the head room and to prevent water droplets from forming inside the igloo, the corners of the blocks may be cut off to form a round and smooth interior.

A shelter may be built over the entrance in the same manner as the igloo, and it should be 1 to 2 meters long and about 1 meter wide and 1 meter high. This shelter provides more protection and warmth inside the igloo and a place to store equipment. Blocks of snow or other material may be placed at both the outside and the inside entrances of the tunnel for more warmth. Make a ventilation hole near the top and at the same height as the fifth or sixth row of snow blocks. The hole can be kept open by placing a ski pole or long stick into it and clearing it out as needed. The igloo is heated with a one-burner stove or with candles.

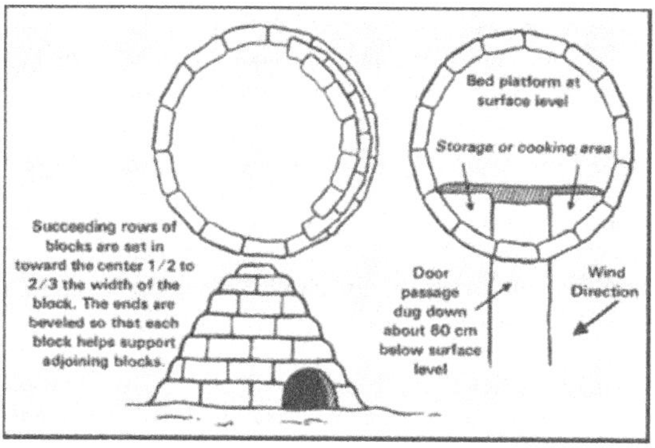

Figure 10-9. Snow house.

10-4. FOOD

The chances of finding different types of food vary in cold-weather areas depending on the time of year and the place. Shores normally are scraped clean of all animals and plants by winter ice. Food can still be found north of the timberline, even when such foods as mice, fish, and grubs are not available.

STORAGE AND PRESERVATION

If a large animal is killed or a large number of small game are found, you should store or preserve some of the meat for future use. Freezing is the best way to preserve fresh meat or fish. It may be necessary to protect your supplies from small animals. This can be done by hanging these supplies about 2 meters from the ground.

FISH

There are few poisonous fish in cold-weather areas, but some fish, like the sculpin, lay poisonous eggs. The black mussel may be poisonous at any season and is as deadly as strychnine. In cold regions do not eat shark meat; it is poisonous. In coastal streams and rivers, salmon moving upstream to spawn may be plentiful; however, their flesh deteriorates as they travel away from the coast, making them unfit to eat except in an emergency. A salmon that has been in fresh water long enough to turn a brilliant red is decaying and may be unfit to eat.

Coastal waters are rich in all seafood. Grayling, trout, white fish, burbot, and pike are common to the lakes, ponds, and arctic coastal plains of North America and Asia. River snails or freshwater periwinkles are plentiful in the rivers, streams, and lakes.

Fish can be speared, shot, netted, hooked, caught by hand, or stunned by a rock or club. Fish hooks can be improvised from pins, pieces of wire, or any other metal object that can be bent into a hook. Pieces of meat, insects, or minnows can be used for bait. Some fish will nibble at any small object that hits the water. Cod will swim up to investigate strips of cloth or bits of metal or bone, and they are easily caught. A net can be made out of stout twine or from the inner strands of parachute suspension lines. These strands can also serve as fishing lines.

LAND ANIMALS

Deer, caribou, reindeer, musk-oxen, moose, elks, mountain sheep, goats, bears, and other such animals are found in most cold-weather areas.

Tundra animals include rabbits, mice, lemmings, ground squirrels, wolves, and foxes. Where there are trees, the porcupine is often encountered. The porcupine can be easily clubbed or shaken out of trees. Pick up a porcupine only after it is dead.

Hunting is better during the early morning or late evening when the animals are moving to and from feeding and bedding grounds and water. Large game such as moose and caribou should be shot just back of the front shoulder or, if the range is short and you are sure of your shot, in the neck. These large animals are fairly easy to stalk and kill, and their skins are also useful for survival. Bears should be shot just forward of the front shoulder to stop them or, if you are sure of your shot, in the neck. To successfully hunt these land animals you should know something of their characteristics:

- Caribou or reindeer are usually very curious. It is possible to attract them near enough for a shot by waving a cloth and moving slowly toward them on hands and knees.
- The technique of imitating a four-legged animal may also cause a wolf to come closer to a hunter.
- Moose may be found in heavy brush or around the shoreline of lakes.
- Mountain goats and sheep are cautious and hard to approach. They can be surprised, however, by moving quietly downwind while they are feeding. If possible, stay on higher ground than they are on.

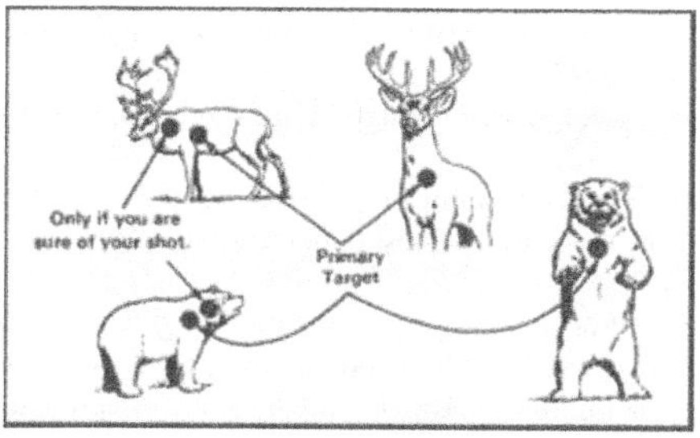

Figure 10-10. Hunting land animals.

- Musk oxen leave cattle-like tracks and droppings. When alarmed, they group together in a circle with their heads facing out and remain in that position unless approached, then one or more bulls may charge.
- Bears are always dangerous. A wounded bear is most dangerous and should not be followed into cover. The polar bear is a tireless, clever hunter with good sight and an extraordinary sense of smell. Be careful that you don't become the hunted instead of the hunter.
- Rabbits often run in circles and return to the same place when they are frightened. A running rabbit can sometimes be stopped by whistling. Snares made of wire and placed on small game trails, arranged so that the animal will place his head through the loop, are an excellent means of catching rabbits.
- The quick-kill method can be used on most small animals and sometimes on larger animals after they have been shot and are on the ground. The method is simple and consists of cutting the animal's throat with a sharp knife. This system not only kills the animal but bleeds him at the same time.

Figure 10-11. Snaring a rabbit.

SEA ANIMALS

During winter and spring, sea mammals — seals, walruses, and polar bears — are found on the frozen pack ice and on floes in open water.

Seals are hard to approach but can be stalked. Keep downwind and avoid sudden moves. A white camouflage suit will help. Advance only when the animal is sleeping. Do not eat the liver of the bearded seal, because its high vitamin A content may cause you to be sick.

Walruses are found on pack ice and must be approached by boat. Because of its large size and boldness, the walrus is one of the most dangerous animals in the arctic.

Polar bears are found in almost all arctic regions but rarely appear on land. Avoid them if possible. If it is necessary to kill one for food, do not eat the liver. The liver is high in vitamin A, which may cause you to be sick. Never eat any bear meat unless it is cooked; it is always diseased.

BIRDS

The breeding ground of many birds is in the arctic. Ducks and geese build their nests near ponds on the coastal plains during the summer and provide an important source of food. Grouse and ptarmigan live in mountainous terrain and brush-covered areas in arctic and subarctic areas. Sea birds may be found on cliffs or small islands on the coast. Sea birds, ravens, and owls are useful for food.

In winter, owls, ravens, grouse, and ptarmigans are the only birds available. Ptarmigans and grouse are easily approached and are very tame. Although they are hard to locate because of the protective coloring, they provide a good source of food because they can be killed with a stone or a club.

PLANTFOOD

Most plants that grow in cold regions can be eaten. The water hemlock and the baneberry are the only poisonous plants. The water hemlock is one of the world's most poisonous plants. It can be recognized by where it grows, always in wet ground, and by the following characteristics: a hollow, partitioned bulb at the base of a hollow stem, spindle-shaped roots, and a foul odor, which is very noticeable in the root and bulb. The water hemlock is abundant in marshes near southern beaches and around marshy lakes in interior river valleys. It is never found on hillsides or dry ground. The baneberry is normally found on hillsides or in forested bog areas. Some mushrooms are poisonous and should not be eaten. Although the cold-region areas have many kinds of berries that are edible, most will not be available in the winter.

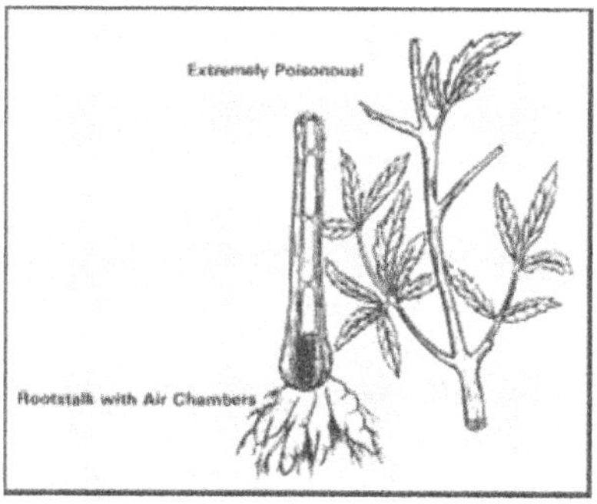

Figure 10-12. Water hemlock.

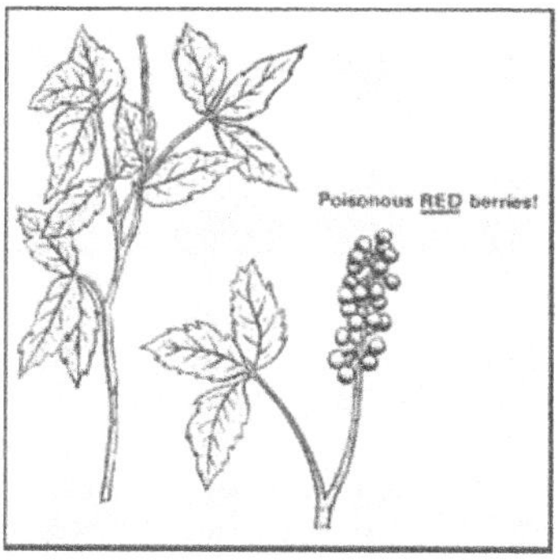

Figure 10-13. Baneberry.

10-5. USEFUL KNOTS

A rope is only a limp coil of hemp or nylon. It is useful only when you are able to attach it to itself or another object. The six basic knots described below will fit all your needs for tying rope together or onto equipment or personnel.

THE HALF HITCH

The half hitch is used to secure the end of a rope to some other object or to secure other knots.

Figure 10-14. Half hitch.

THE OVERHAND KNOT

The overhand knot is used to make a knotted rope for a handline or to temporarily whip the end of a rope to keep it from unraveling.

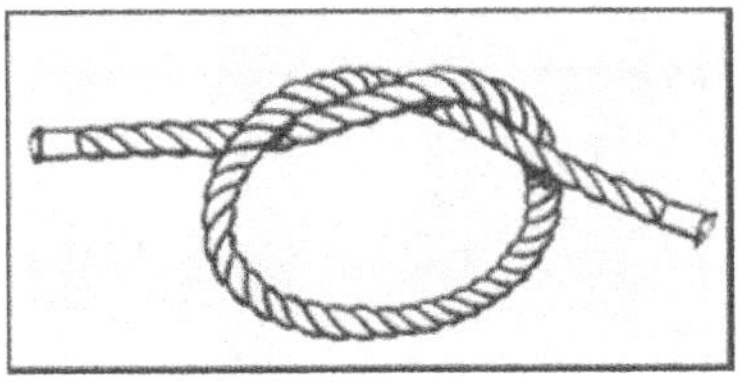

Figure 10-15. Overhand.

THE SQUARE KNOT

The square knot is used to tie the ends of two ropes of equal width together. The knot should be secured by tying a half hitch on each side of the knot. If properly tied, the square knot will not slip or jam. Be careful to follow Figure 10-16 or you may end up with a "granny knot," which can slip and jam.

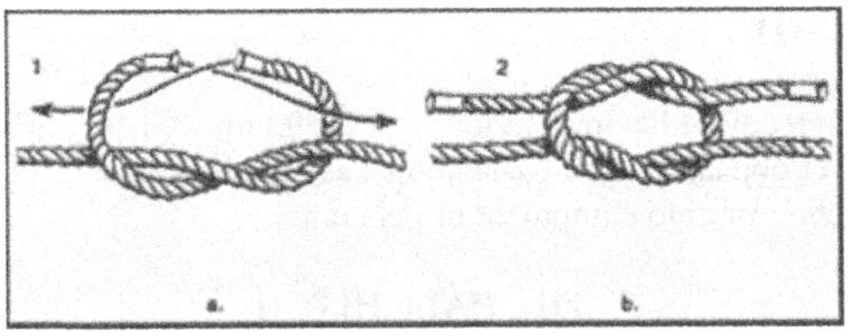

Figure 10-16. Square.

THE CLOVE HITCH

The clove hitch is the most useful of the hitches. It can be used to tie the end of a rope to a post or eye, or it can be used to secure the middle of a rope without using the ends.

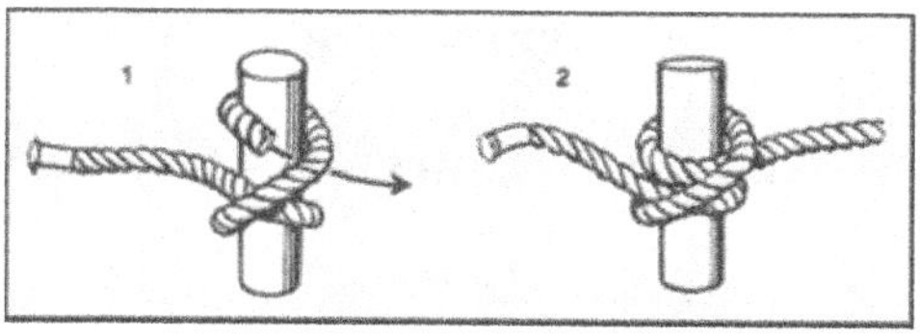

Figure 10-17. Clove hitch.

THE DOUBLE SHEET BEND

The double sheet bend is used for tying two ropes together of different width.

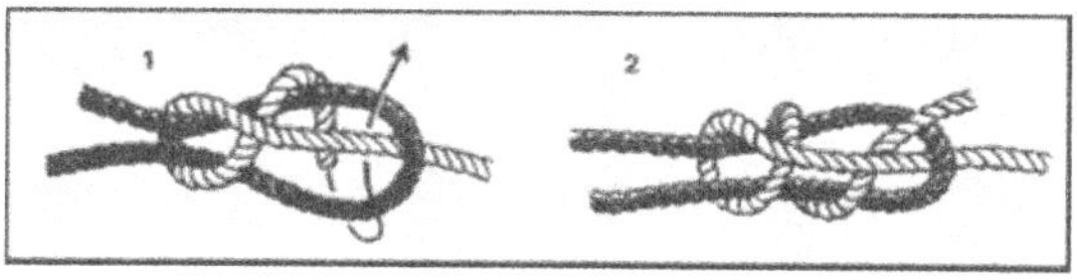

Figure 10-18. Double sheet bend.

THE BOWLINE

The bowline is used to form a loop on the end of the rope that will not slip.

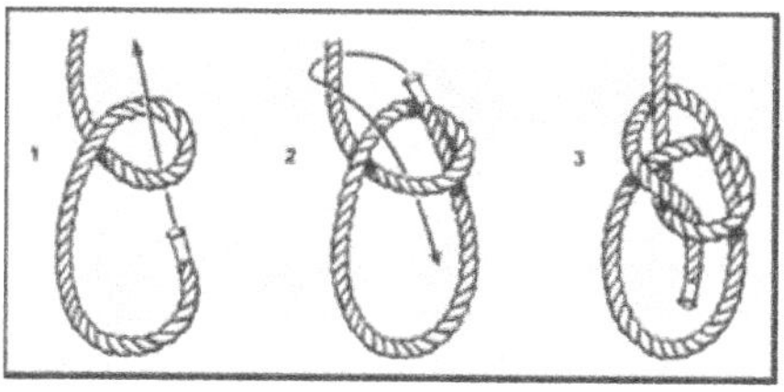

Figure 10-19. Bowline.

Remember the qualities of a good knot: it is easy to tie, it does not jam and become hard to untie, and it does not slip when weight is put on it.

APPENDIX B

Approximate Measurement Conversion

Factors

Multiply	*By*	*To Obtain*
Centimeters	.03	Feet
Centimeters	.4	Inches
Meters	3.3	Feet
Meters	40.0	Inches
Meters	1.1	Yards
Kilometers	.6	Miles
Knots	1.2	MPH
Inches	2.5	Centimeters
Feet	30.0	Centimeters
Yards	.9	Meters
Miles	1.6	Kilometers
MPH	.9	Knots

For ease in conversion the above factors have been rounded off.

References

Suggested Readings

DA Pam 20-201	Military Improvisations During the Russian Campaign
DA Pam 20-203	Russian Combat Methods in World War II
DA Pam 20-231	Combat in Russian Forests and Swamps
DA Pam 20-234	Operations of Encircled Forces: German Experiences in Russia
DA Pam 20-236	Night Combat
DA Pam 20-240	Rear Area Security in Russia: The Soviet Second Front Behind the German Lines
DA Pam 20-269	Small Unit Actions During the German Campaign in Russia
DA Pam 20-271	The German Northern Theater of Operations, 1940-1945
DA Pam 20-290	Terrain Factors in the Russian Campaign
DA Pam 20-292	Warfare in the Far North
FM 9-207	Operation and Maintenance of Ordnance Materiel in Cold Weather (0° to -65°F)
FM 31-70	Basic Cold Weather Manual
FM 31-71	Northern Operations
FM 90-6	Mountain Operations
FM 90-11 (TBP)	Winter Operations
TC 90-11-1	Military Skiing

"Perhaps the most valuable result of all education is the ability to make yourself do the things you have to do when it ought to be done, whether you like it or not."

Hosley

Glossary

ALICE	all-purpose lightweight individual carrying equipment
C	Celsius
CLP	cleaner lubricant preservative
ECWCS	extended cold-weather clothing system
F	Fahrenheit
FM	field manual
GP	general purpose
SFP	synthetic fiber pile
SOP	standing operating procedure
TBP	to be published
VB	vapor barrier

Made in the USA
Monee, IL
05 November 2026

67980372R00105